Anika Bargsten

Characterization of trace gas fluxes in and above soil

Anika Bargsten

Characterization of trace gas fluxes in and above soil

reactive and non reactive trace gases

Südwestdeutscher Verlag für Hochschulschriften

Impressum/Imprint (nur für Deutschland/ only for Germany)
Bibliografische Information der Deutschen Nationalbibliothek: Die Deutsche Nationalbibliothek verzeichnet diese Publikation in der Deutschen Nationalbibliografie; detaillierte bibliografische Daten sind im Internet über http://dnb.d-nb.de abrufbar.

Alle in diesem Buch genannten Marken und Produktnamen unterliegen warenzeichen-, marken- oder patentrechtlichem Schutz bzw. sind Warenzeichen oder eingetragene Warenzeichen der jeweiligen Inhaber. Die Wiedergabe von Marken, Produktnamen, Gebrauchsnamen, Handelsnamen, Warenbezeichnungen u.s.w. in diesem Werk berechtigt auch ohne besondere Kennzeichnung nicht zu der Annahme, dass solche Namen im Sinne der Warenzeichen- und Markenschutzgesetzgebung als frei zu betrachten wären und daher von jedermann benutzt werden dürften.

Verlag: Südwestdeutscher Verlag für Hochschulschriften GmbH & Co. KG
Dudweiler Landstr. 99, 66123 Saarbrücken, Deutschland
Telefon +49 681 37 20 271-1, Telefax +49 681 37 20 271-0
Email: info@svh-verlag.de
Zugl.: Bayreuth, Uni, Diss, 2010

Herstellung in Deutschland:
Schaltungsdienst Lange o.H.G., Berlin
Books on Demand GmbH, Norderstedt
Reha GmbH, Saarbrücken
Amazon Distribution GmbH, Leipzig
ISBN: 978-3-8381-2553-4

Imprint (only for USA, GB)
Bibliographic information published by the Deutsche Nationalbibliothek: The Deutsche Nationalbibliothek lists this publication in the Deutsche Nationalbibliografie; detailed bibliographic data are available in the Internet at http://dnb.d-nb.de.

Any brand names and product names mentioned in this book are subject to trademark, brand or patent protection and are trademarks or registered trademarks of their respective holders. The use of brand names, product names, common names, trade names, product descriptions etc. even without a particular marking in this works is in no way to be construed to mean that such names may be regarded as unrestricted in respect of trademark and brand protection legislation and could thus be used by anyone.

Publisher: Südwestdeutscher Verlag für Hochschulschriften GmbH & Co. KG
Dudweiler Landstr. 99, 66123 Saarbrücken, Germany
Phone +49 681 37 20 271-1, Fax +49 681 37 20 271-0
Email: info@svh-verlag.de

Printed in the U.S.A.
Printed in the U.K. by (see last page)
ISBN: 978-3-8381-2553-4

Copyright © 2011 by the author and Südwestdeutscher Verlag für Hochschulschriften GmbH & Co. KG and licensors
All rights reserved. Saarbrücken 2011

Characterization of reactive and non reactive trace gas fluxes in and above soil

A dissertation submitted to the
FACULTY OF BIOLOGY, CHEMISTRY AND GEOSCIENCES
AT THE UNIVERSITY OF BAYREUTH

for the degree of
DR. RER. NAT.

presented by
ANIKA BARGSTEN

Dipl. Geographin

born in Buxtehude

2010

Characterization of reactive and non reactive trace gas fluxes in and above soil

Supervisor: Prof. Dr. Bernd Huwe

Die vorliegende Arbeit wurde in der Zeit von Januar 2007 bis Juni 2010 am Lehrstuhl für Bodenphysik der Universität Bayreuth unter der Betreuung von Herrn Prof. Dr. Bernd Huwe und am Max-Planck-Institut für Chemie in Mainz unter der Betreuung von Prof. Dr. Franz X. Meixner angefertigt.

Vollständiger Abdruck der von der Fakultät für Biologie, Chemie und Geowissenschaften der Universität Bayreuth genehmigten Dissertation zur Erlangung des akademischen Grades Doktor der Naturwissenschaften (Dr. rer. nat.).

Amtierender Dekan: Prof. Dr. Stephan Clemens
Tag des Einreichens der Dissertation: 24.Juni 2010
Tag des wissenschaftlichen Kolloquiums: 26.Oktober 2010

Prüfungsausschuß:
Prof. Dr. Franz X. Meixner (Erstgutachter)
Prof. Dr. Andreas Held (Zweitgutachter)
Prof. Dr. Thomas Foken (Vorsitzender)
Prof. Dr. Harold Drake
Prof. Dr. Bernd Huwe

"Overall, our understanding of the nitrogen cycle and the development of effective policies to reduce inadvertent losses of anthropogenic nitrogen to the environment is analogous to our understanding of the carbon cycle the late 1960s.

Humans are adding nitrogen to the earth's surface; we do not know where it all goes, but we do know that increasing concentrations of nitrogen in unexpected places will cause significant environmental damage (…)."

William H. Schlesinger (2009)

Summary

Nitrogen is one of the most important compounds on earth. All organisms need nitrogen to live and grow. Even the majority (78.08%) of the atmosphere (and so the air we breathe) is dinitrogen. Over the last century, human activities have dramatically increased emissions and removal of nitrogen to the global atmosphere by as much as three to five fold. Nitrous oxide is the fourth largest single contributor to positive radiative forcing, and serves as the only long-lived atmospheric tracer of human perturbations of the global nitrogen cycle. Nitrogen oxides belong to the so called indirect greenhouse gases. These indirect greenhouse gases control the abundances of direct greenhouse gases through atmospheric chemistry and contribute on this way to the greenhouse effect. For a better understanding of these feedback mechanisms it is necessary to know the source strength of nitric oxide and nitrous oxide. Thus, the knowledge about exchange processes of nitrogen is of interest and importance for scientist and policy makers, likewise.

This thesis contributes the understanding of processes in the nitrogen cycle. The thesis is addressed on nitric and nitrous oxide emissions. Nitric oxide emissions were measured on soil samples using an automated laboratory system. Nitrous oxide emissions were measured directly on the field site using a closed chamber technique.

The laboratory measurements were compared with field measurements of NO (modified Bowen ratio method) at a grass land site. The field NO fluxes were always around 1.8 ng m^{-2} s^{-1} while the laboratory derived NO fluxes were between 2.1 and 5,2 ng m^{-2} s^{-1}. The agreement between the two data sets is considered to be quite good. The laboratory derived NO fluxes exceeded the field NO fluxes by a factor of 1.5 to 2.5.

Most studies of nitric oxide (NO) emission potentials up to now have investigated mineral soil layers only. In this thesis soil organic matter was sampled for laboratory measurements under different understory types (moss, grass, spruce, blueberries) in a humid mountainous Norway spruce forest plantation in the Fichtelgebirge (Germany). In this thesis the response of net potential NO fluxes on physical and chemical soil conditions (water content and temperature, bulk density, particle density, pH, C/N ratio, organic C, soil ammonium, soil nitrate) was determined. Net potential NO fluxes (in terms of mass of N) from soil samples taken under

Summary

the different understories ranged from 1.7 - 9.8 ng m^{-2} s^{-1} (soil sampled under grass and moss cover), 55.4 - 59.3 ng m^{-2} s^{-1} (soil sampled under spruce cover), and 43.7 - 114.6 ng m^{-2} s^{-1} (soil sampled under blueberry cover) at optimum water content and a soil temperature of 10°C. Effects of soil physical and chemical characteristics on the net potential NO flux were statistically significant (0.01 probability level) only for NH_4^+. Therefore, as an alternative explanation for the differences in soil biogenic NO emission we consider more biological factors like understory vegetation type, amount of roots, and degree of mycorrhization; they provide a potential explanation of the observed differences of net potential NO fluxes.

Also, soil nitrous oxide (N_2O) emissions in an unmanaged, old growth beech forest in the Hainich National Park, Germany, were measured at 15 plots over a one-year period (November 2005 to November 2006). The annual field N_2O flux rate was 0.46±0.32 kg ha^{-1} yr^{-1}. The N_2O emissions showed a background emission pattern with two event based N_2O peaks. A correlation analysis showed that the distance between plots (up to 380 m) was secondary for their flux correlations. Annual N_2O fluxes obtained from a standard model (Forest-DNDC) parameterized with soil parameters as well as daily temperature and precipitation substantially overestimated the actual field N_2O fluxes and also did not describe their actual temporal and spatial variabilities. Temporal variability was described well by the model only at plots with higher soil organic carbon and the modelled N_2O fluxes increased during freezing periods only were soil organic carbon was larger than 0.06 kg^{-1} C kg. The results indicate that the natural background of nitrous oxide emissions may be lower than previously thought and also lower than assumed in standard modelling. This suggests a higher anthropogenic contribution to N_2O emissions.

Zusammenfassung

Stickstoff ist eines der wichtigsten Elemente auf der Erde. Alle Organismen benötigen Stickstoff zum Leben und Wachsen. Der Großteil (78,08%) der Erdatmosphäre - und daher auch die Luft, die wir atmen - besteht aus molekularem Stickstoff (N_2). Im letzten Jahrhundert haben menschliche Aktivitäten die Stickstofffreisetzung und den Stickstoffabbau in der Atmosphäre drei- bis vierfach erhöht. Distickstoffoxid (N_2O, Lachgas) liefert den viertgrößten Beitrag zur Änderung der globalen Strahlungsbilanz und ist außerdem der einzige langlebige atmosphärische Tracer, der den Einfluss des Menschen auf den globalen Stickstoffkreislauf widerspiegelt. Stickstoffmonoxid (NO) gehört zu den sogenannten indirekten Treibhausgasen. Indirekte Treibhausgase beeinflussen den Gehalt von direkten Treibhausgasen durch ihren Anteil an chemischen Reaktionen in der Atmosphäre und tragen auf diesen Weg ihren Teil zum Treibhauseffekt bei. Um diese Prozesse besser zu verstehen ist es wichtig die Quellenstärke von Stickstoffmonoxid und Distickstoffoxid zu kennen. Ebenso ist das Wissen und Verständnis um die Austauschprozesse von Stickstoff sowohl für Wissenschaftler als auch für politische Entscheidungsträger von Bedeutung.

Diese Doktorarbeit möchte einen Beitrag zum besseren Verständnis des Stickstoffkreislaufes leisten. Die Arbeit befasst sich mit den natürlichen Emissionen von NO und N_2O. NO-Emissionen wurden mit einem automatischen Laborsystem an Bodenproben gemessen. Emissionen von N_2O wurden hingegen direkt auf der Untersuchungsfläche im Freiland bestimmt. Hierfür wurden statische Kammern verwendet.

Die NO-Labormessungen wurden mit den NO-Feldmessungen (Modifizierte Bowen-Verhältnis Methode) verglichen. Die Feldmessungen wurden dabei über Grasland durchgeführt. Die in situ bestimmten NO-Flüsse schwankten die gesamte Messperiode hindurch um 1,8 ng m^{-2} s^{-1}. Die im Labor gemessenen NO-Flüsse bewegten sich dagegen zwischen 2,1 und 5,2 ng m^{-2} s^{-1}. Beide Datensätze zeigten eine gute Übereinstimmung. Die im Labor bestimmten NO-Flüsse waren lediglich 1,5 bis 2,5 mal höher als die NO-Flüsse, die in situ bestimmt wurden.

Zusammenfassung

Die meisten Studien, die sich bis heute mit NO-Emissionen befasst haben, betrachteten die mineralischen Bodenhorizonte. Für diese Arbeit wurden hingegen auch organische Horizonte beprobt. Die Proben wurden unter verschiedenen Unterwuchstypen (Moos, Gras, Fichten, Blaubeeren) eines humiden Fichtenwaldes im Fichtelgebirge (Deutschland) genommen und anschließend im Labor untersucht. Die Untersuchungen umfassten das Verhalten der potentiellen Netto-NO-Flüsse unter Berücksichtigung bodenphysikalischer und -chemischer Parameter (Wassergehalt, Bodentemperatur, Lagerungsdichte, Partikeldichte, pH-Wert, C/N-Verhältnis, organischer Kohlenstoff, Ammonium, Nitrat). Die potentiellen Netto-NO-Flüsse (in Einheiten von N) der Bodenproben unterschieden sich je nach Unterwuchstypen bei optimalem Wassergehalt und einer Bodentemperatur von 10°C. Für Bodenproben der von Moos und Gras bewachsenen Flächen lagen die potentiellen Netto-NO-Flüsse zwischen 1,7 – 9,8 ng m^{-2} s^{-1}. Dahingehen wurden für Bodenproben von Flächen, die mit Fichten bewachsen waren, Werte zwischen 55,4 – 59,3 ng m^{-2} s^{-1} gemessen. Für Flächen, die mit Blaubeeren bewachsen waren, variierten die NO-Flüsse der Bodenproben zwischen 43,7 – 114,6 ng m^{-2} s^{-1}. Ein Zusammenhang zwischen den physikalischen und chemischen Bodenparametern und dem potentiellen Netto-NO-Fluss konnte nur für NH_4^+ (0.01 Signifikanzlevel) gefunden werden. Daher wurde nach alternativen Erklärungen für diese Unterschiede in den biogenen NO-Flüssen gesucht. Gründe für die Unterschiede in den potentiellen Netto-NO-Flüssen könnte die Vegetation, die Menge der Wurzel oder der Anteil an Mycorrhiza sein.

Die N_2O-Emissionen eines altbestehenden Buchenwaldes des Nationalparks Hainich (Deutschland) wurden gemessen um die Quellenstärke eines über lange Zeit nicht bewirtschaften Waldes zu bestimmen. Die Messungen wurden an 15 Messpunkten über die Dauer von einem Jahr (November 2005 bis November 2006) durchgeführt. Der Jahreswert des N_2O-Flusses betrug für diesen Zeitraum 0.46±0.32 kg ha^{-1} a^{-1}. Die N_2O-Emissionen zeigten "Background-Emissionen" mit zwei eventbasierten Peaks. Durch eine Korrelationsanalyse konnte gezeigt werden, dass die Distanz zwischen den Messpunkten (bis zu 380 m) für die Korrelation der Flüsse zweitrangig ist. Zusätzlich wurde ein Jahreswert für den N_2O-Fluss mit einem Standardmodell (Forest-DNDC) berechnet. Zur Parametrisierung des Modells dienten Bodenparameter sowie die Tagestemperatur und der tägliche Niederschlag. Der mit dem Modell berechnete N_2O-Fluss überschätzt den tatsächlich auf der

Zusammenfassung

Untersuchungsfläche gemessenen Fluss. Außerdem wird die zeitliche und räumliche Variabilität nicht korrekt wiedergegeben. Der zeitliche Verlauf wurde von dem Modell nur bei Messpunkten mit einem hohen organischen Kohlenstoffgehalt korrekt dargestellt. Ebenso stieg der modellierte N_2O-Fluss während Frostperioden nur an Messpunkten mit einem organischen Kohlenstoffgehalt über 0.06 kg^{-1} C kg an. Die Ergebnisse zeigten, dass der natürliche Hintergrund von N_2O-Emissionen niedriger sein kann als angenommen und auch niedriger als der mit einem Standardmodell berechnete Jahreswert. Diese deutet einen höheren anthropogenen Beitrag zu den N_2O-Emissionen an als angenommen.

List of manuscripts

The present cumulative thesis consists of three manuscripts. The first manuscript has been published. The second manuscript has been submitted for publication and the third manuscript will be submitted soon.

Published manuscript

A. Bargsten, E. Falge, B. Huwe and F.X. Meixner: Laboratory measurements of nitric oxide release from forest soil with a thick organic layer under different understory types, Biogeoscience, 7, 1425 – 1441, 2010.

Submitted manuscript

J.-C. Mayer, A. Bargsten, U. Rummel, F.X. Meixner and T. Foken: Distributed modified Bowen ratio method for surface layer fluxes of reactive and non-reactive trace gases, Agricultural and Forest Meteorology, accepted, 2010.

Manuscript in preparation

A. Bargsten, M. Timme, S. Glatzel and H. Jungkunst: Low nitrous oxide fluxes in an unmanaged old growth beech forest, European Journal of Soil Science, to be submitted, 2010.

Danksagung

Die Zeit der Doktorarbeit war eine interessante und schöne, aber auch teilweise schwere Phase in meinem Leben. Glücklicherweise gab es Personen, die mich in dieser Zeit unterstützt, mich aufgebaut und meine Freizeit mit mir verbracht haben oder auf andere Weise für mich da waren.

- Als erstes gilt mein Dank Prof. Dr. Franz X. Meixner, der mich in seine Arbeitsgruppe am Max Planck Institut für Chemie in Mainz aufnahm und mir ein sehr interessantes Thema zur Verfügung stellte.
- Ebenso möchte ich meinem Doktorvater, Prof. Dr. Bernd Huwe von der Universität Bayreuth, danken. Er nahm mich als Doktorandin auf und ermöglichte mir so diese Promotion.
- Dankbar für die Finanzierung dieser Arbeit bin ich der Max Planck Gesellschaft (MPG) und der Deutschen Forschungsgemeinschaft (DFG).
- Mein besonderer Dank gilt Dr. Eva Falge, die sich immer Zeit für mich genommen hat und mir mit Rat und Tat beiseite stand. Ohne Eva hätte ich wahrscheinlich nie ein Licht am Ende des Tunnels gesehen.
- Danken möchte ich auch Dr. Hermann Jungkunst und Dr. Marc Timme für hilfreiche Diskussionen und die Zeit, die sie sich genommen haben.
- Meinen Kollegen vom Max Planck Institut möchte ich für die tolle Zusammenarbeit und zahlreiche Kaffeerunden und danken. Mein besonderer Dank gilt Alex und Jens.
- Großen Dank auch der super IOP-1-Crew! Ohne euch wäre die Zeit im Fichtelgebirge sehr trostlos und langweilig gewesen.
- Mein Dank gilt auch den Mitarbeitern der Werkstatt, der Elektronik und des Grafikbüros.
- Vielen Dank auch an meine Boulder- und Klettercrew für die Stunden, die wir beim Training oder am Fels verbracht haben und natürlich für eure Freundschaft.
- Nicht zuletzt möchte ich meiner Mutter danken, die mir immer zugehört hat und für mich da war.
- Danke!

Contents

Summary	I
Zusammenfassung	III
List of manuscripts	VI
Acknowledgements	VII
Contents	VIII
Synthesis	1
1. Introduction	1
1.1 N-Cycle	1
1.1.1 Nitric oxide and nitrous oxide	2
1.2 Nitrogen in soil	4
1.2.1 Nitrification	5
1.2.2 Denitrification	6
1.2.3 Chemodenitrification	7
1.3 Factors controlling biogenic NO and N_2O emissions from soil	7
1.3.1 Soil temperature	8
1.3.2 Soil moisture	9
1.3.3 Other controlling factors	9
1.4 Modeling NO and N_2O emissions	10
1.4.1 Black box models	11
1.4.2 White box models	12
1.5 Objects of this thesis	15
2. Experiments	16
2.1 EGER/Fichtelgebirge	16

Contents

    2.2 LIBRETTO/Brandenburg   17

    2.3 Carbon storage in an unused beech forest in the Hainich national park - Differentiation of the soil carbon source and sink considering land use history /Thuringia   17

    2.4 Laboratory setup   18

3. Results   20

    3.1 Comparison between laboratory and filed measurements of NO   20

    3.2 NO fluxes in the Fichtelgebirge   21

    3.3 N_2O fluxes in the Hainich   24

    3.4 Modeling N_2O fluxes   26

4. Conclusion   26

5. Reverences   30

Appendix A

Individual contribution to the publications   **35**

Appendix B

Laboratory measurements of nitric oxide release from forest soil with a thick organic layer under different understory types   **39**

Abstract   40

1 Introduction   40

2 Material and methods   42

    2.1 Sample site   42

    2.2 Soil sampling and preparation   43

    2.3 Soil physical and chemical characterization   45

    2.4 Laboratory setup   45

    2.5 Calculation and fitting the net NO release rate   47

	Contents
2.6 NO compensation point mixing ratio	49
2.7 Net potential NO flux	49
2.8 Calculation of the Q_{10} value	50
2.9 Effective diffusion of NO in soil air	50
2.10 Error estimation of NO release measurements	52
3 Results	54
3.1 Net NO release rates	54
3.2 NO production rates, NO consumption coefficients, and NO compensation point mixing ratio	56
3.3 Net potential NO fluxes	59
3.4 Temperature dependence (Q_{10} value)	61
3.5 Chemical and physical soil parameters	63
4 Discussion	64
4.1 Comparison with other studies	64
4.2 Influence of soil chemical parameters on net potential NO flux	69
4.3 Influence of understory type on net potential NO flux	70
5 Conclusion	73
Acknowledgements	74
References	75

Appendix C

Distributed Modified Bowen Ratio Method for Surface Layer Fluxes of reactive and non-reactive Trace Gases — **83**

Abstract	84
Keywords	84
1 Introduction	84
2 Material and methods	87

2.1 Site and setup	87
2.2 Quality control and gap filling	91
2.2.1 Reference data	91
2.2.2 Profile station data	93
2.3 Distributed modified Bowen ratio (DMBR)	93
2.4 Laboratory setup	95
2.5 Boundary layer budget method	97
2.6 Characteristic time scale	97
3 Results and Discussion	101
3.1 Horizontal homogeneity	101
3.2 Characteristic time scale	102
3.3 Thermodynamic conditions of exchange	104
3.4 Trace gases – mixing ratios	105
3.4.1 Advection	109
3.5 Trace gases – fluxes	110
3.5.1 Comparison of NO fluxes: field vs. laboratory	113
3.5.2 Comparison of methods: DMBR vs. NBLB (nocturnal boundary layer budget)	114
4 Conclusion	120
Acknowledgements	121
References	122

Appendix D

Low nitrous oxide emissions in an unmanaged old growth beech forest   **127**

Abstract	128
1 Introduction	128

	Contents
2 Materials and methods	130
2.1 Study site	130
2.2 Field measurements and N_2O flux analysis	130
2.3 Forest-DNDC	131
2.4 Climate and soil parameter	134
2.5 Calculation of the quality of simulation	134
2.6 Statistical analysis	135
3 Results	136
3.1 Field N_2O fluxes	136
3.2 Soil climate	139
3.3 Modelled N_2O fluxes	140
3.4 Quality of simulation	141
3.5 Spatial distribution of N_2O fluxes	142
3.6 Spatial correlation of N_2O fluxes	143
3.7 Physical and chemical soil parameters	144
4 Discussion	146
4.1 Comparison with other studies	146
4.2 Measured vs. modelled N_2O fluxes	148
4.3 Spatial variability of N_2O fluxes	150
4.4 Spatial correlation of N_2O fluxes	150
5 Conclusion	151
Acknowledgement	151
References	152

Synthesis

1 Introduction

1.1 Nitrogen Cycle

The nitrogen cycle represents one of the most important nutrient cycles found in terrestrial ecosystems. All organisms need nitrogen (N) to live and grow. Even the majority (78.08%) of the atmosphere (and so the air we breathe) is dinitrogen (N_2), but most of the nitrogen in the atmosphere is unavailable for organisms. This is because N_2 is relatively inert. In order to use nitrogen, organisms must first convert N_2 to a more "available" form such as ammonium (NH_4^+) and nitrate (NO_3^-). Because of the inert nature of N_2, biologically available nitrogen is often in short supply in natural ecosystems, limiting plant growth and biomass accumulation.

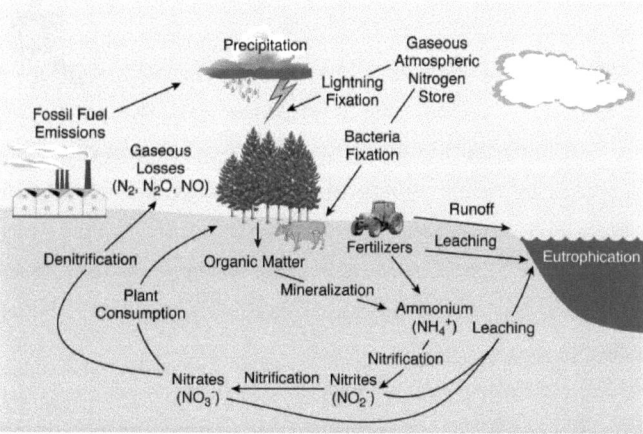

Figure 1: Nitrogen cycle (modified after Brady and Weil (2007))

Nitrogen exists in inorganic and organic forms and there are many different oxidation states. Fig. 1 displays the movement of nitrogen between the atmosphere, biosphere, and geosphere in different forms. The nitrogen cycle consists of various reservoirs and five main processes which are responsible for the exchange among them. Those main processes are nitrogen fixation, nitrogen uptake, nitrogen mineralization, nitrification and denitrification. In the nitrogen transformation microorganisms play an important role. The nitrogen cycle is affected by environmental factors like temperature, moisture, resource availability, or anthropogenic influence.

1.1.1 Nitric oxide and nitrous oxide

Nitric oxide (NO) and nitrous oxide (N_2O) play an important role in the chemistry of the troposphere and stratosphere (Crutzen, 1979). N_2O is one of the natural components of the Earth's atmosphere and is a long-living greenhouse gas. The atmospheric concentration of N_2O has increased from about 270 ppb during the pre-industrial era to 319 ppb in 2005 resulting in a total source strength of about 17.7 Tg-N yr^{-1}(Denman et al., 2007). N_2O is (after conversion to NO) also responsible for destruction of the stratospheric ozone (O_3). The soil is the dominant source of atmospheric N_2O contributing about 57% of the total annual global emission (Denman et al., 2007). N_2O in soils is mainly produced and consumed by two microbial processes, nitrification and denitrification. Natural soils contribute 6.0 Tg-N yr^{-1} and agricultural soils 4.2 Tg-N yr^{-1}(Denman et al., 2007). Nevertheless, soils can also act as a sink for atmospheric N_2O (Chapuis-Lardy et al., 2007).

Atmospheric NO is a reactive trace gas which has a short atmospheric lifetime of hours or days, and is also known as an indirect greenhouse gas. Due to the fast chemical interconversion with nitrogen dioxide (NO_2), which typically occurs within seconds to minutes, both species are commonly referred to as the single quantity NO_x (NO_x = NO + NO_2). Through atmospheric chemistry NO_x affects the abundance of direct greenhouse gases (Prather and Ehhalt, 2001). NO is removed from the atmosphere mostly through oxidation processes that involve reactions with hydroxyl radicals (OH·) and ozone. Hence, NO has a significant influence on the oxidation capacity of the troposphere, especially due to its catalytic behaviour in the photochemical formation of O_3 (Crutzen, 1979).

NO_x catalyses tropospheric O_3 formation through a sequence of reactions. When mixtures of NO and O_3 are exposed to ultraviolet light, an equilibrium is established in which NO reacts with O_3 to form NO_2 and oxygen (O_2) and vice versa.

$NO + O_3 \rightarrow NO_2 + O_2$ (1.1)

$NO_2 + h\nu \rightarrow NO + O\cdot$ (1.2)

$O\cdot + O_2 \rightarrow O_3$ (1.3)

Table 1: Overview of global sources (Tg-N yr^{-1}) of NO$_x$ and N$_2$O. Values are from the Third Assessment Report (according to Denman et al., 2007).

Sources	NO$_x$	N$_2$O[a]
Anthropogenic sources		
Fossil fuel combustion and industrial processes	33	1.3/0.7
Aircraft	0.7	-
Agriculture	2.3	6.3/2.9
Biomass and biofuel burning	7.1	0.5
Total, anthropogenic	**43.1**	**8.1/4.1**
Natural sources		
Soils under natural vegetation	3.3	6.0/6.6
Oceans	-	3.0/3.6
Lightning	5	-
Atmospheric chemistry	<0.5	0.6
Total, natural	**8.8**	**9.6/10.8**
Total, all sources	**51.9**	**17.7/14.9**

[a] a single value indicates agreement between the sources and methodologies of the different studies.

The absorption of ultraviolet radiation protects the biosphere from harmful radiation. In the process electronically excited O(^{1}D) atoms are generated:

$$O_3 + h\nu \rightarrow O(^1D) + O_2 \qquad (\leq 310 nm) \tag{1.4}$$

These excited O(^{1}D) atoms can react with water vapour (H$_2$O) to form very reactive OH· radicals (Crutzen, 1979):

$$O(^1D) + H_2O \rightarrow 2\ OH· \tag{1.5}$$

Fossil fuel combustion and biomass burning are the main sources of NO_x (see Tab. 1.1). However, in the less industrial European countries, soils have been estimated to contribute between 24 and 62% of the total annual emission (Skiba et al., 1997).

1.2 Nitrogen in soils

Nitrogen occurs only to a very small part in the parent rock material. Therefore, organisms need other sources of N. Nitrogen is present in soil in organic and inorganic forms. Organic compounds are e.g. amino acids, or large complex molecules that are quite resistant to microbial decay. The most resistant of these soil organic materials are typically referred to as humus. Inorganic forms of N are nitrate (NO_3^-), nitrite (NO_2^-), ammonium (NH_4^+), and ammonia (NH_3). NO_3^- and NH_4^+ are taken up by plants, whereas NO_2^- and NH_3 are toxic to plants.

N_2 is the most abundant form of N in the biosphere but is unusable for most organisms, including plants. Biological N_2 fixation by microorganisms is necessary, whereby N_2 is transformed to organic N. This is the dominant process by which N first enters soil biological pools (Robertson and Groffman, 2007) (see Fig. 1). High energy natural events such as lightning can also lead to N fixation. The organic N will be microbiologically transformed to inorganic forms of N. This process is termed mineralization. Common organic N substances are: soil humus, plant leafs and roots and manure based fertilizer. Generally, a complex and large molecule containing N is broken down into a simpler and smaller molecule and then into NH_4^+ that can be taken up again by plants or other organisms (Robertson and Groffman, 2007). Sometimes this process is referred to in two steps with the first step termed aminization and the second step ammonification. A lot of different types of microorganisms can perform these reactions, some can do both steps while others can only perform one reaction (Scheffer and Schachtschabel, 2002). If plant detritus is rich in N, mineralization, or N release, proceeds. If plant detritus is low in N, microorganisms take up mineral N out of the soil solution (process of immobilization, which is the uptake or assimilation of inorganic forms of N by microbes and other soil heterotrophs).

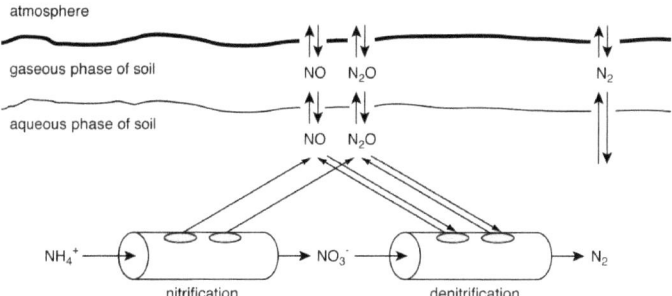

Figure 2: Diagram of the hole-in-the-pipe conceptual model (revised from Davidson 1991). Soil emissions of NO and N_2O are regulated at two levels: First, the rate of nitrogen cycling through ecosystems, which is symbolized by the amount of nitrogen flowing through the pipes, affects total emissions of NO and N_2O; second, soil water content and other factors affect the ratio of N_2O: NO emissions, symbolized by the relative sizes of the holes through which nitric oxide and nitrous oxide "leak."

Mineralization results in an increase, while immobilization results in a decrease, in plant-available forms of N in soil. Usually, NH_4^+ is seen as the immediate product of mineralization (Robertson and Groffman, 2007). However, mineralization and immobilization occur at the same time within relatively small volumes of soil. While one group of microorganisms might consume a detritus rich in N (mineralization), another group might consume detritus low in N (immobilization). Mineralization and immobilization can be carried out by a wide range of organisms (aerobes, anaerobes, fungi, bacteria). Both processes are strongly influenced by climate, soil properties and soil management.

The mineralization is followed by the process of nitrification and then the process of denitrification (see Fig. 2). It is generally accepted that these two biological processes are the principal sources of NO and N_2O emissions (Williams et al., 1992). NO and N_2O are intermediates in the nitrification and denitrification pathway. Generally most NO is produced by nitrifiers and most N_2O is produced by denitrifiers. However, it is not possible to generalise this because each soil has a different behaviour. It is also possible that soils act as a sink for both gases (Conrad, 2002, 1996).

1.2.1 Nitrification

Nitrification is an important part of the nitrogen cycle in soils. It is a mainly aerobic process in which ammonium is oxidised to NO_2^- and to NO_3^- (Eq. 1.6). NO_3^- is the final product of the

Synthesis

nitrification process. Most nitrifying bacteria are autotrophic and linked to O_2, but when O_2 is limited the nitrifying bacteria can use NO_2^- as an electron acceptor and reduce it to NO and N_2O (Bollmann and Conrad, 1998). Therefore, NO and N_2O can be released during nitrification. Nitrification is a two-step process in which two different groups of microorganisms are involved. The first step is carried out by *Nitroso*-bacteria (ammonia oxidizers) and the second step by *Nitro*-bacteria (nitrite oxidizers).

$NH_4^+ \rightarrow NO_2^- \rightarrow NO_3^-$ (1.6)

The *Nitroso*-bacteria oxidise NH_4^+ via hydroxylamine (NH_2OH) to NO_2^-. Intermediary compounds formed during the oxidation of NH_2OH to NO_2^- can result in the formation of NO (see Fig. 3) (Robertson and Groffman, 2007). *Nitroso*-bacteria seem also able to produce NO via NO_2^- reduction, which results in the production of N_2O. NO_2^- reduction occurs when *Nitroso*-bacteria use NO_2^- as an electron acceptor when O_2 is limiting (denitrifying nitrifiers). In most soils NO_2^- produced by *Nitroso*-bacteria does not accumulate as *Nitro*-bacteria quickly oxidize NO_2^- to NO_3^-.

Nitrification is also observed by heterotrophic bacteria and fungi, while two pathways for heterotrophic ammonia oxidation exist. The first pathway is similar to the pathway of *Nitroso*-bacteria, as the nitrifying bacteria have similar ammonia- and hydroxylamine-oxidizing enzymes. The second heterotrophic pathway seems to be limited to fungi.

However, autotrophic nitrification appears to be the dominant process of NO production in most soils (Conrad, 1996). The magnitude of nitrification is influenced by many factors, which have a direct or indirect influence on the nitrification process. For example, the optimum temperature for nitrification in soils is between 25 and 35°C. However, also at temperatures around 0°C nitrification occurs.

1.2.2 Denitrification

Denitrification is the reduction of NO_3^- to the N gases NO, N_2O, and N_2 (Eq. 1.7). Denitrifiers are aerobic microorganisms which can switch to anaerobic denitrification in the absence of O_2. Denitrification generally occurs under anaerobic conditions and requires nitrate and microbially available organic carbon (organic matter). The temperatures in which denitrification occurs range from 5 to 75°C (Scheffer and Schachtschabel, 2002).

$NO_3^- \rightarrow NO_2^- \rightarrow NO \rightarrow N_2O \rightarrow N_2$ (1.7)

Denitrifiers use NO_3^- rather than O_2 as a terminal electron acceptor during respiration. As O_2 is the more efficient electron acceptor, most denitrifiers only carry out denitrification when O_2 is unavailable. Such a situation occurs in most soils after rainfall, as the soil pores become water-saturated and the diffusion of O_2 through the soil declines. Hence, high soil water contents or limited aeration are important for denitrification. Typically, denitrification starts to occur at a water filled pore space (WFPS) of 60% (see Fig. 3), where little NO is released from soil to the atmosphere (Conrad, 1996). Hence, denitrification is the major source of atmospheric N_2O. Both gases, NO and N_2O, are produced as intermediates during the denitrification process. Thus, the denitrification process is commonly associated with the loss of soil nitrogen to the atmosphere.

Not only denitrifying microorganisms reduce NO_3^-, there are also several other biological processes that reduce NO_3^- and consequently produce NO and N_2O (Robertson and Groffman, 2007).

1.2.3 Chemodenitrification

Chemodenitrification occurs when NO_2^- in soil reacts to form N_2, NO or N_2O. Generally, chemodenitrification occurs in acidic soils (pH<5). There are several aerobic pathways for chemodenitrification. However, in most ecosystems chemodenitrification is only a minor pathway for N loss (Meixner & Yang, 2006).

1.3 Factors controlling biogenic NO and N_2O emission from soil

The individual factors that regulate N_2O and NO production and consumption are e.g. temperature, moisture, soil bulk density, soil texture, soil pH, soil nutrients, plants, ambient concentration of NO and N_2O. The same environmental factors affect both NO and N_2O (Davidson et al., 2000). Both nitrification and denitrification have their own set of optimum conditions. As a result, one process may be the primary N_2O producer in one set of field conditions, but as soil conditions change, another process may predominate. The complexity of the interacting factors important to the different processes obviously makes a simple description of N_2O and NO production difficult (Mosier et al., 1983). The following sections will give an overview of the main controlling factors, soil temperature and soil moisture, and a short outline of other controlling factors.

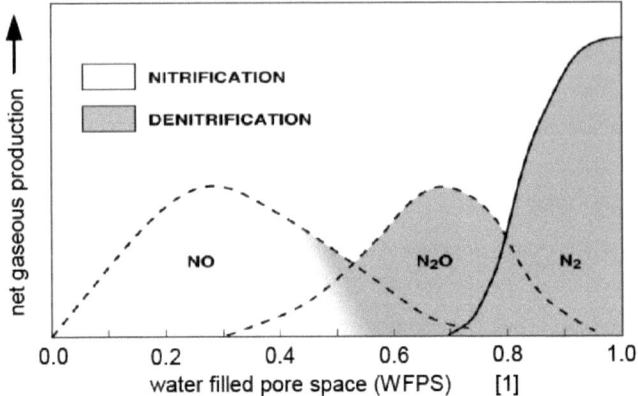

Figure 3: Proposed relative contributions of nitrification and denitrification to emissions of NO and N_2O as a function of water filled pore space (taken from Meixner & Yang, 2006). Shapes of the curves, the curve heights, WFPS optima, and the inflection points were largely educated guesses based on the limited field and laboratory data available

1.3.1 Soil temperature

Soil temperature is one of the main factors controlling the exchange of NO and N_2O between soil and atmosphere. Many studies have shown that NO and N_2O emission increases with increasing temperature (e.g. Otter et al., 1999; Johansson, 1984; Meixner and Yang, 2006; Smith et al., 1998; Smith et al., 2003). This is due to the fact that both NO and N_2O production are microbial processes. These microbial processes are influenced by temperature according to the Arrhenius equation (Winkler et al., 1996). Generally, enzymatic processes increase exponentially with temperature as long as other factors, like soil moisture or soil nutrients, are not limiting (Ludwig et al., 2001). The outcome of this could be a diurnal or seasonal variation of NO and N_2O release (e.g. Baumgärtner and Conrad, 1992; Smith et al., 1998; Brumme and Borken, 2009; Flessa et al., 2002; Ludwig et al., 2001; Christensen, 1983). The temperature response is usually expressed in terms of the Q_{10} value. The Q_{10} value gives the increase of the exchange rate by a temperature increase of 10°C. Microbial processes usually show a Q_{10} value of 2 (Smith et al., 2003). However, Q_{10} values up to 10 are no rarity for N_2O production in soils (Smith et al., 2003).

1.3.2 Soil moisture

In numerous studies, soil moisture was found to be one of the most important controlling factors (e.g. Feig et al., 2008; Meixner and Yang, 2006; Bollmann and Conrad, 1998; Pilegaard et al., 1999; Davidson et al., 2000). Soil moisture affects the diffusion of oxygen (O_2) in soil as well as the substrate supply of microorganisms. When the soil moisture is very low the diffusion of O_2 to the microorganisms is high, but the supply with substrate is low and vice versa (Skopp et al., 1990; Davidson et al., 2000). Hence, NO emissions are highest at intermediate soil moisture. This intermediate soil moisture range varies from soil to soil. In the literature, optimum soil moistures for NO emissions range between 10% and 70% (Yang and Meixner, 1997; Pilegaard et al., 1999; Bargsten et al., 2010). N_2O emissions are also highest at intermediate soil moistures. When water content is greater than field capacity, N_2O gets reduced to N_2 (Bremner and Blackmer, 1979). However, the production of NO is generally faster than the corresponding production of N_2O. This is reflected in the ratios in which the two gases are emitted from soils (Smith et al., 2003). Generally, more NO is released at lower soil water contents than N_2O and vice versa. This is due to the fact that NO is mainly released by nitrification and N_2O is mainly released by denitrification (Bollmann and Conrad, 1998) (see Fig. 3).

1.3.3 Other controlling factors

Temperature und soil moisture are the main important parameters controlling the exchange of NO and N_2O. However, there are also a lot other factors influencing the exchange of both gases.

Soil bulk density (BD) is an important factor controlling the NO and N_2O exchange because the compaction of the soil influences the diffusion of O_2 to the microorganisms and additionally the release of NO and N_2O is hindered. Generally, NO and N_2O exchange decreases with increasing BD.

The effect of *soil texture* on NO and N_2O emission results from the physical variations of air and water properties. Water infiltration rate and gas diffusion rate are affected by soil texture and hence affect NO and N_2O emissions. Coarse textured soil has a relatively smaller total pore space, compared to a fine textured soil. At an identical soil moisture (mass of water per mass of soil), a coarse textured soil would be relatively wet compared to a fine textured soil.

It is difficult to give a value for the optimum soil pH for NO and N_2O emissions. The optimum soil pH for NO and N_2O emissions via nitrification and denitrification varies with species and age of the organisms and NO_3 concentration. However, most nitrifiers have a pH optimum between pH 5.5 and pH 8.0 and most denitrifiers between pH 6.0 and pH 8.0 (Pathak, 1999; Scheffer and Schachtschabel, 2002). Although the processes are favoured at slightly alkaline soil pH levels, they also take place in acidic soils. Chemodenitrification occurs mostly in acid soils (soil pH < 5.5).

The availability of *soil nutrients*, especially NH_4^+ and NO_3^-, affects the emissions of NO and N_2O because these compounds serve as a substrate for nitrifying and denitrifying microorganisms. Skiba et al. (1994) and Ludwig and Meixner(1994) showed that differences in the NO_3^- content in soil accounted for much of the variance in the observed NO emissions.

Plants influence the emission of NO and N_2O by affecting nitrate and carbon content of the soil as well as partial pressure of O_2. Plants can directly affect the availability of NO_3^- through uptake and assimilation making it unavailable to denitrifiers. However, mineralization of roots and other plant material to NH_4^+ and nitrification of NH_4^+ to NO_3^- can potentially provide more NO_3^- for denitrification and conversely immobilization can reduce NO_3^- in the soil.

The *ambient (atmospheric) concentration* of NO and N_2O determines whether a given soil acts as a sink or source for NO and N_2O. That is due to the fact that NO and N_2O production and consumption occur simultaneously in the soil. Therefore, fluxes of NO and N_2O are bi-directional. The concentration at which the production and consumption of NO and N_2O is equal is termed compensation point mixing ratio. However, in most studies a compensation point mixing ratio above the average ambient air concentration was observed – that means the soil acts as a source.

1.4 Modelling NO and N_2O emissions

Estimates of the contribution of temperate forest ecosystems to atmospheric NO and N_2O derived by field measurements have a high degree of uncertainty. These uncertainties are mainly due to:

- the limited number of field measurements,

- the limited temporal resolution of data sets,
- the limited information on the impact of anthropogenic influence,
- and the limited knowledge about the effects of different forests types.

Field measurements alone are not enough to significantly reduce these uncertainties associated with global estimates of NO and N_2O source strengths of temporal forests. The most promising strategy to overcome these problems is the development of models.

Nowadays, models are used in numerous studies to estimate emissions of NO or N_2O from soils. A lot of research groups have tried to model NO and N_2O emissions from soils, from relatively small scales up to global scales (Beirle et al., 2004b; Butterbach-Bahl et al., 2001; Steinkamp et al., 2009; Saggar et al., 2004; Beirle et al., 2004a; Martin et al., 2003; Martin and Asner, 2005). Generally, there are different kinds of models, which can be roughly categorized into "Black box models", "White box models" and hybrid types of both ("Grey box models").

1.4.1 Black box models

Black box models are used where the response of a system is not broken down into its underlying mechanisms. It is represented by an empirical description or set of transfer parameters that relate the output of the model to a set of inputs. In a pure black box model the internal workings of a device are not described, and the model simply solves a numerical problem without reference to any underlying processes. Exemples for Black box models are:

- *Empirical models* were used for the first approaches to model NO and N_2O emissions (Williams et al., 1992). The model relates the NO emissions to soil temperature and a biome fitting parameter (emission factor) which is supposed to be representative of an ecosystem. Empirical models operate by assigning various ecosystems an emission factor. These emission factors are often modified according to parameters like soil moisture or soil temperature. Even though process-based models (biogeochemical models) have been developed, empirical models are widely used to estimate NO and N_2O emissions due to the rudimentary knowledge of NO and N_2O producing processes.

Synthesis

- *Statistical models* use relationships between measured NO or N_2O fluxes and physical and chemical parameters of soils to estimate the emission of NO and N_2O from soils. A statistical model is designed to fit to an existing data set. Then it can be used to predict the NO and N_2O emission from measured or modelled environmental parameters. Statistical models were used e.g. by Yan et al. (2005) or Delon et al. (2007).

1.4.2 White box models

White box models are the most detailed types of models. To a white box model a full set of priori information is ready. When having a scientific theoretic foundation of a system, it is possible to provide the model with a priori-knowledge usually given in the form of ordinary differential equations describing how different aspects of the system changes over time. Therefore, white box models are close to the full description of a real system. Examples for white box models which are used in this thesis are:

- A *local model* to determine NO fluxes *based on soil measurements* was developed by Galbally and Johansson (1989). The model assumes that the net exchange of NO can be determined in terms of NO production, NO consumption, and NO diffusion through the soil. For this, NO production and NO consumption are determined with soil samples in the laboratory. The Galbally and Johansson model was validated by various studies (Remde et al., 1993; van Dijk et al., 2002; Otter et al., 1999; Mayer et al., 2010; Meixner et al., 1997; Ludwig et al., 2001).

  The laboratory measurements of NO production and NO consumption under varying temperature and ambient NO concentration are used to determine the fluxes of NO from soils (Galbally and Johansson, 1989). Hence, the measurements are conducted over a wide range of soil moistures (0-100% WFPS) and varying soil temperatures. Then, the NO flux can be estimated for that particular type of soil as a function of the soil moisture and temperature. Using land use distributions and measurements or estimates of the soil temperature and moisture, the NO fluxes can be up-scaled to a larger region (Kirkman et al., 2001; Yu et al., 2008; van Dijk et al., 2002; Feig et al., 2009).

- *Biogeochemical models* simulate the movement of nutrients through ecosystems by looking at the important processes, such as the rate of decomposition, the rate of nitrification and denitrification etc. There are a lot of different process-based models, such as CENTURY (Parton et al., 1996), DNDC (Li et al., 1992), PnET-N-DNDC (Li et al., 2000), ExpertN(Baldioli et al., 1994), NASA CASA (Potter et al., 1996).

We briefly refer to the model of Li et al. (2000) as an example of biogeochemical models. The challenges of modelling NO and N_2O emissions byPnET-N-DNDC attribute to three reasons (Li et al., 2000) (see Fig. 4):

- NO and N_2O are multisource gases, as there are at least three sources: nitrification, denitrification and chemodenitrification. These three processes are so different in their dynamics and kinetics that, when they are mixed together, the pattern of NO and N_2O fluxes is very complex.
- The reactions are driven by a number of forces including soil environmental parameters (e.g. temperature, moisture...) and ecological drivers (e.g. climate, soil properties...). Any change in the combination of the forces will alter the magnitude and/or pattern of NO and N_2O fluxes.
- NO and N_2O are intermediates of nitrification and denitrification. This means that the fluxes of NO and N_2O are determined by the kinetics of production, consumption, and diffusion of the gases in the sequential biochemical reactions.

The PnET-N-DNDC is a fusion of new developments with three existing models (the Photosynthesis-Evapotranspiration (PnET) model, the Denitrification-Decomposition (DNDC) model, and the nitrification model). Table 2 will give an overview of the prediction of the three models.

Field and laboratory studies have shown a complex picture of soil NO and N_2O emissions from various sources which are directly influenced by a number of soil environmental factors (temperature, moisture, pH, and substrate availability). These soil environmental factors are controlled by several ecological drivers, such as climate, soil physical properties, vegetation, and anthropogenic activities. Two components were constructed in the PnET-N-DNDC model to reflect the links between the ecological drivers, the soil environmental factors, and NO and N_2O fluxes (Li et al., 2000).

Synthesis

Table 2: Overview of the three models integrated in the PnET-N-DNDC model.

Model	PnET	DNDC	nitrification model
Prediction	forest photosynthesis, respiration, organic carbon production and allocation, litter production	soil decomposition, denitrification	nitrifier growth/death rates, nitrification rate, nitrification-induced NO and N_2O production

The first component contains three interacting submodels to quantify impacts of ecological drivers on the soil environmental factors (see Fig. 4). The soil climate submodel simulates soil temperature, soil moisture, and redox potential profiles based on daily climate data, soil physical parameters, soil water status, thermal impacts of plants, and soil respiration. The forest growth submodel simulates forest growth driven by solar radiation, temperature, water stress, and N stress, and passes the litter production, water and N demands, and root respiration to the soil climate submodel or the decomposition submodel. The decomposition submodel tracks concentrations of substrates, like dissolved organic carbon, NH_4^+ and NO_3^-, based on climate, soil properties, and management measures (Li et al., 2000).

The second component consists of two submodels (see Fig. 4). This component predicts impacts of the soil environmental factors on nitrification and denitrification. The nitrification submodel predicts NO and N_2O production by tracking growth and death of nitrifiers under aerobic conditions. The denitrification submodel simulates growth and death of denitrifiers, substrate consumption, and gas diffusion under anaerobic conditions. Fluxes of NO and N_2O are a result of competition among the kinetics of production, consumption, and diffusion of the two gases in the soil. The five interacting submodels link the ecological drivers to the NO and N_2O emissions. The soil, climate and decomposition algorithms were adopted from the DNDC model and the forest growth submodel was adopted from the PnET model (Li et al., 2000).

Synthesis

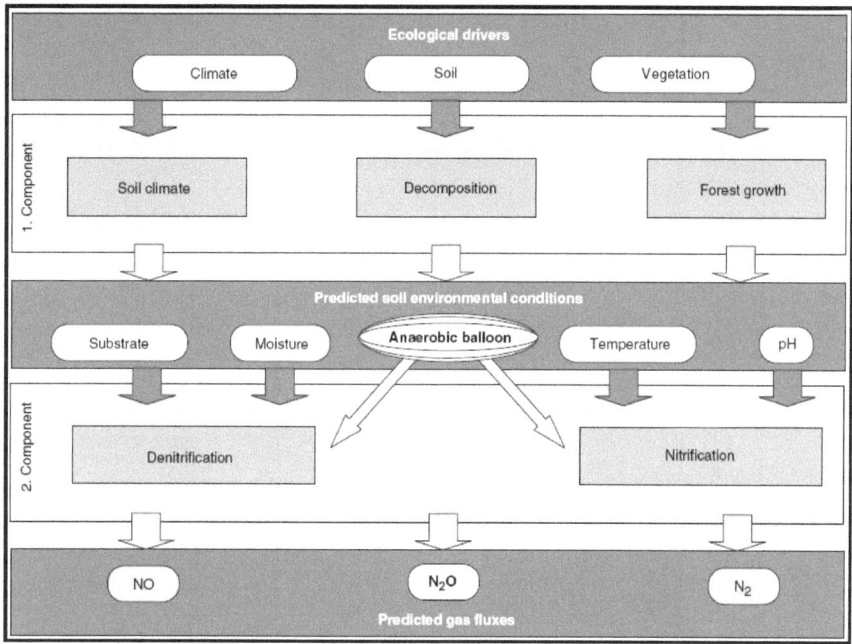

Figure 4: Schematic overview of the PnET-N-DNDC model: components, submodels, ecological drivers and environmental conditions. Figure is taken from Kiese et al. (2005).

1.5 Objectives of this thesis

The aim of this thesis focuses on small scale NO and N_2O fluxes measured in the laboratory and in the field as well as on the comparison with modelled N_2O fluxes. The thesis is structured in four main parts.

1. The validation of the laboratory measurements of NO through a comparison between laboratory measurements and the modified Bowen ratio technique at a grass land site in Brandenburg, Germany.

2. Laboratory measurements of NO emissions from organic layers from the Weidenbrunnen site, Germany, with respect to the spatial variability of NO fluxes. The main focus is to investigate the influence of different understory types in a Norway spruce forest on net potential NO fluxes as well as the relationship with physical and chemical soil parameters.

Synthesis

3. Closed chamber measurements of N_2O over a period of one year at the Hainich research site, Germany. The measurements were performed with a high spatial distribution to investigate the influence of the distances between the different plots. Also, the influence of physical and chemical soil parameters on N_2O fluxes was studied.

4. The comparison between modelled and field N_2O fluxes measured at the Hainich research site. For modelling N_2O fluxes the process-based model Forest-DNDC was used.

2 Experiments

The results presented in this thesis are based on datasets obtained during three projects in which the author participated. Measurements were performed at the field site as well as in the laboratory.

2.1 EGER / Fichtelgebirge

Field and laboratory measurements were conducted within the framework of the project EGER (ExchanGE processes in mountainous Regions (www.bayceer.uni-bayreuth.de/vp_eger/)) at the research site "Weidenbrunnen" (50°08'N, 11°52'E, 774 m a.s.l.) located in the Fichtelgebirge Mountains, NE Bavaria, Germany. The site is mainly covered by 55 year old Norway spruce (*Piceaabies*) with significant variability in the understory. There are four different main understory types: moss, grass (*Deschampsiaflexuosa* and *Calamagrostisvillosa*), blueberries (*Vacciniummyrtillus*), and young spruce which cover 45, 19, 7 and 13%, respectively, of the total surface area of the Weidenbrunnen site (Behrendt, 2009).The mean annual air temperature of the Weidenbrunnen site is 5.3°C, mean annual soil temperature is 6.3°C, and mean annual precipitation is approximately 1160 mm (1971 - 2000; Foken, 2003; Falge et al., 2003). The soil type was classified as cambicpodzol over granite (Subke et al., 2003), and the texture is sandy loam to loam, with relatively high clay content in the Bh horizon. The mineral soil is characterised by low pH values (<4). The soil litter and the organic horizon had a thickness between 5 and 9 cm (Behrendt, 2009). The organic layer is classified as a moder consisting of Oi, Oe, and Oa horizons. More details concerning the site can be found in Gerstberger et al. (2004).

Measurements were performed from August to October 2007 and soil samples for laboratory analyses were taken in September 2008.

2.2 LIBRETTO / Brandenburg

The LIBRETTO (LIndenBergREacTive Trace gas prOfiles) campaign took place in late summer 2006, from 01 August 2006 until 31 August 2006 at the Falkenberg Boundary Layer Field Site of the Meteorological Observatory Lindenberg (Richard-Aßmann Observatory) (Beyrich and Adam, 2007). The field site is located at 52° 10' 01" N, 14° 07' 27" E, 73 m a.s.l.. The main vegetation species are perennial ryegrass (*Loliumperenne*), red fescue (*Festucarubra*), dandelion (*Leontodonautumnalis, Taraxacumofficinale*), bromegrass (*Bromushordeaceus*), and clover (*Trifoliumpratense, Trifoliumrepens*). The meadow is mowed regularly in order to keep the mean vegetation height below 20 cm (Beyrich and Adam, 2007). The measuring site comprises one 99 m and one 10 m high profile mast (air temperature (T), relative humidity (rH), wind speed (u) and wind direction), two identical setups for the measurement of the net radiation flux, two stations for the measurement of turbulent fluxes of momentum, sensible and latent heat (further on referred to as the EC stations), and a sub-site to monitor physical soil quantities (soil temperatures, soil heat flux and soil moisture). A SODAR-RASS system completes the permanent setup of the Falkenberg site.

2.3 Carbon storage in an unused beech forest in the Hainich national park - Differentiation of the soil carbon source and sink considering land use history / Thuringia

Field measurements were conducted within the framework of the project"Carbon storage in an unused beech forest in the Hainich national park - Differentiation of the soil carbon source and sink considering land use history" at a research site located in the Hainich National Park (51°04'46''N, 10°27'08''E, 440 m a.s.l.), Thuringia, Germany. The Hainich National Park was established in 1997 to protect one of the largest beech forests in Central Europe. Due to a unique history as a military base for more than 60 years prior to 1997, a large part of the forest has been taken out of management and developed basically undisturbed. In the centuries before, the forest at the Hainich research site was used by the local village population as a coppice with standard systems and therefore has not been exposed to clearcut. As a consequence, the trees cover a wide range of age classes with a maximum up to 250 years. The forest is dominated by beech (*Fagussylvatica*, 65%). The above-ground stem carbon pool is about 130 t C ha^{-1} (Mund, personal communication). Maximum tree height varies between 30 and 35 m with a maximum leaf area index (LAI) of 5.0 m^2 m^{-2}. The long-term mean annual air temperature is 7.5 - 8°C and the mean annual precipitation is 750 – 800 mm.

Measurements were performed from November 2005 to November 2006.

2.4 Laboratory setup

Net NO release rates from soil samples taken at the Lindenberg and Weidenbrunnen site were determined using an automated laboratory system. A detailed description of our experimental setup is given in van Dijk and Meixner(2001); here we give only a short description of the most recent state of the setup.

Pressurized air is passed through a pure air generator (PAG 003, ECOPHYSICS, Switzerland) to provide dry and NO-free air. This NO-free air supplied five Plexiglas cuvettes (four incubation cuvettes and one empty reference cuvette). The volume of each cuvette was $9.7*10^{-4}$ m^3 (0.97 l)) and each was flushed with a continuous flow of $4.2*10^{-5}$ m^3 s^{-1} (2.5 l min^{-1}) of dry NO-free air, as controlled by five mass flow controllers (MFC, Mass-Flo®, 5000 sccm range, MKS instruments, USA), one for each cuvette. The headspace volume of each cuvette is well mixed by a teflonized micro-fan (Micronel®, USA). The outlet of each cuvette was connected to a switching valve. Every two minutes one cuvette was switched to be the "active" cuvette (i.e., connected to the analyzers, while the remaining four cuvettes were still purged), so that all five cuvettes were measured within 10 minutes. The valves provided necessary sample air to a chemiluminescence detector, NO-analyser (Model 42i Trace Level, Thermo Electron Corporation, USA; detection limit: 250 ppt (3σ)) and a CO_2-/H_2O-analyzer (Li-cor 840, Licor, USA). Instead of ambient air we operated the NO-analyser with pure oxygen (O_2) to obtain a better accuracy and precision of the NO mixing ratio measurements, particularly at low mixing ratios.

The NO-analyser was calibrated using a gas phase titration unit (GPT, 146 C Dynamic Gas Calibrator, Thermo Electron Corporation, USA). For operating the GPT we used NO-free air from the PAG 003 and an NO gas standard (5.02 ppm NO, Air Liquide, Germany). The determination of the soil NO compensation mixing ratio (Conrad, 1994) requires the flushing of incubated soil samples with enhanced NO mixing ratios (resulting in reduced or even negative net NO release rates, i.e. NO uptake by the soil). Hence, NO standard gas (200 ppm NO, Air Liquide, Germany) was diluted into the air flow from the PAG 003 via a mass flow controller (Flow EL, Bronkhorst, Germany).

All connections and tubes consisted of polytetrafluorethylene (PTFE). A homebuilt control unit (V25) was controlling the entire laboratory system and was also used, in combination with a computer, for data acquisition.

To determine the temperature response of the net NO release we performed a total of four experiments, each on another sub-sample of the original understory soil sample. The sub-samples were identically pre-treated. Incubations were at 10°C and 20°C, corresponding flushing was either with dry, NO-free air, or with air containing a high NO concentration (soil samples from the Lindenberg site were measured with air containing 50 ppb NO and soil samples from the Weidenbrunnen site were measured with air containing 133 ppb NO). Since every experiment begins with a wetted soil sample and the flushing air is completely dry, the gravimetric water content (θ) of the samples declines during each experiment as evaporating water leaves the cuvette with the flushing air flow. Gravimetric soil moisture content was measured by tracking the loss of water vapour throughout the measurement period and relating this temporal integral to the gravimetric soil moisture content observed at the start and end of the measurement period. This procedure provides us the response of the net NO release rates over the entire range of gravimetric soil moisture.

For measuring the Lindenberg soil samples the laboratory system was run with Nafiondriers. The purpose of the reverse Nafion driers is to keep the humidity of the chambers headspace air high, and hence to slow the dehydration of the soil, allowing the microbes in the soil time to equilibrate to changes in the soil moisture content (Feig et al. 2008). The Weidenbrunnen samples consist on organic material so the samples natural dry slowly.

3 Results

3.1 Comparison between laboratory and field measurements of NO

To build a basis for further measurements and data evaluation, the first part of this thesis is a comparison between laboratory NO fluxes and field NO fluxes. The micrometeorological Distributed Modified Bowen Ratio (DMBR) method was compared with laboratory parameterizations based on the analysis of soil samples. The NO fluxes derived in the laboratory depend on soil temperature and soil moisture measured during the LIBRETTO campaign.

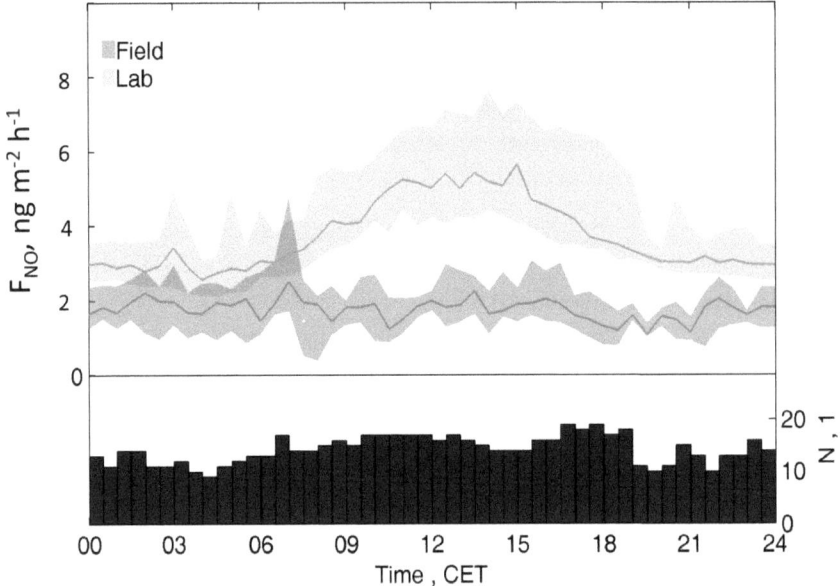

Figure 5: Median diel courses of NO flux from field measurements (green) and from up-scaled (laboratory derived) net potential NO fluxes (grey); up-scaling was achieved with field data of soil moisture and soil surface temperature. Straight lines represent the medians of NO fluxes, while color shaded areas represent their corresponding inter-quartile ranges; blue bars at the bottom indicate the number of data points available for calculation of medians and inter-quartile ranges. Figure is taken from Mayer et al. (accepted).

The obtained time series of laboratory derived NO fluxes was then converted into a median diel cycle, according to the median diel cycle of NO fluxes derived by the DMBR method.

Synthesis

The resulting range of the NO flux is shown together next to the field data in Figure 5. The field NO fluxes were always around 1.8 ng m^{-2} s^{-1} while the laboratory derived NO fluxes were between 2.1 and 5.2 ng m^{-2} s^{-1}. The large scatter in the DMBR results originates from the limitation of data validity to Damköhler numbers less than 0.25. The laboratory NO fluxes exceeded the field NO fluxes by a factor of 1.5 to 2.5. While a considerable, diel amplitude was observed for the laboratory NO flux, the field NO flux remained almost constant around 1.8 ng m^{-2} s^{-1}. Acknowledging the very different nature of both methods to derive NO fluxes, the agreement between the two data sets is considered to be quite good.

The laboratory NO flux is only valid directly at the surface of the soil (cf. Galbally and Johansson, 1989), while the field NO flux is attributed to 0.55 m above ground. Because turbulence is very weak close to the ground, the residence time of NO in the layer between soil surface (0 m above ground) and 0.55 m above ground might be long (compared to the characteristic time of the reaction of NO with O_3 (see section 1.1.1). Therefore, a considerable vertical divergence of the NO flux can occur, which would reduce the NO flux with increasing distance from the soil surface to 0.55 m above ground. Considering this, a factor of two between laboratory data and field data is a good match.

3.2 NO fluxes in the Fichtelgebirge

Laboratory incubation and flushing experiments were performed using a customized chamber technique to determine the response of net potential NO fluxes to physical and chemical soil conditions (e.g. temperature, moisture, bulk density, soil pH). Soil samples for laboratory measurements were taken under different understory types (moss, grass, spruce and blueberry) from the organic layer of the mountainous Norway spruce plantation in the Fichtelgebirge. Net NO release rates showed highest values at soil samples taken under spruce and blueberry cover and lowest values at soil samples taken under moss cover (Fig. 6). Furthermore, soil samples taken under moss cover showed negative values by fumigating with $m_{NO,ref}$=133 ppb that is probably due to the low compensation point mixing ratios of this understory type. The NO compensation point mixing ratios ($m_{NO,comp}$) at 1.0±0.1 gravimetric soil moisture, which is at the end of gravimetric soil moisture observed at the sample site (Behrendt, 2009), varied over a wide range. A low $m_{NO,comp}$ (38 ppb) was observed for soil samples taken under moss cover. For soil samples taken under grass cover a $m_{NO,comp}$ of

94 ppb was determined. Soil samples taken under spruce and blueberry cover showed highest $m_{NO,comp}$ with 518 and 389 ppb.

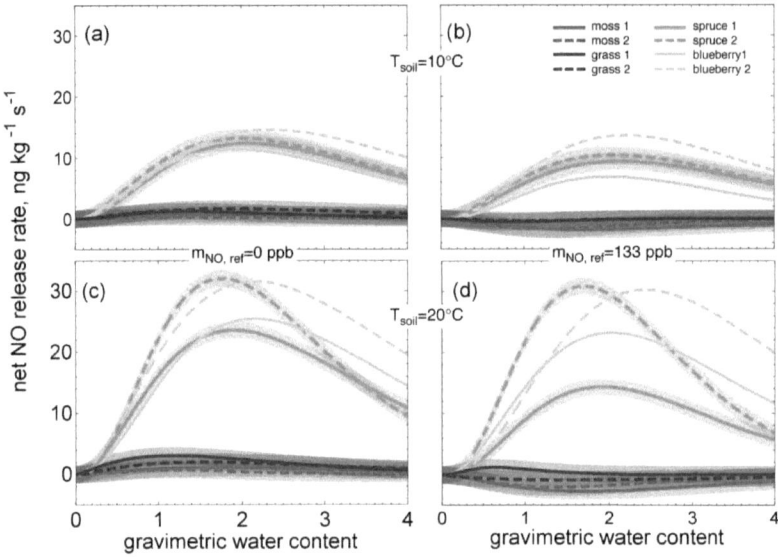

Figure 6. Net NO release rates fitted through experimental results at (a) $T_{soil}=10°C$ and $m_{NO,ref}=0$ ppb, (b) $T_{soil}=10°C$ and $m_{NO,ref}=133$ ppb, (c) $T_{soil}=20°C$ and $m_{NO,ref}=0$ ppb NO and (d) $T_{soil}=20°C$ and $m_{NO,ref}=133$ ppb (all expressed in terms of mass of nitrogen). The transparent bands are the prediction bands of each line (95% confidence level). Figure taken from Bargsten et al. (2010).

The net potential NO fluxes derived from measured net NO release rates from the different soil samples were also lowest for soil samples taken under moss and grass cover (see Fig. 7). It is remarkable that the net potential NO fluxes from soil samples taken under spruce and blueberry cover were approximately 10-fold higher than net potential NO fluxes from soil samples taken under moss and grass cover (note different scales of y-axes in Fig. 7).

Optimum net potential NO fluxes measured at two different soil temperatures (10°C and 20°C) allowed to estimate Q_{10} values for each soil sample of the Weidenbrunnen site. For one of the soil samples taken under spruce cover the lowest Q_{10} value (0.92) was derived while one soil sample taken under blueberry cover showed the highest Q_{10} value of 3.04. These values are in the range of Q_{10} values observed in other studies (Feig et al., 2008; van Dijk et al., 2002; Kirkman et al., 2002).

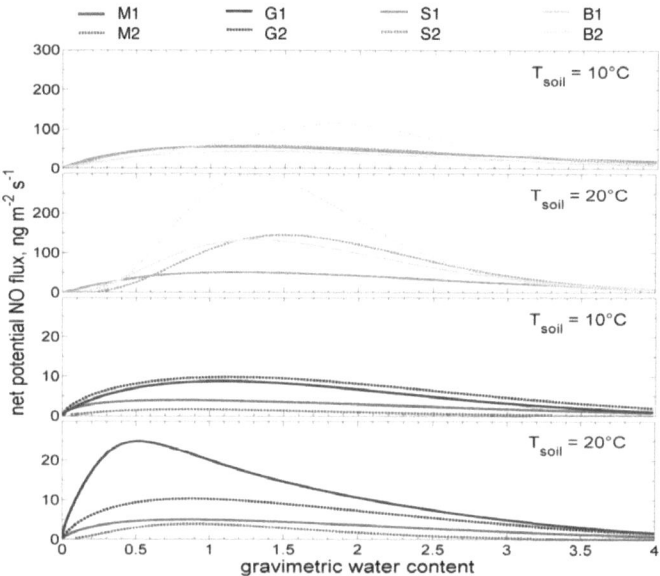

Figure 7: Net potential NO flux (all expressed in terms of mass of nitrogen) at 10°C and 20°C from soil samples taken under moss, spruce and blueberry covered patches (note different scales of the y-axes). Figure taken from Bargsten et al. (2010).

Through a Pearson-product-moment-analysis a significant negative correlation (probability level of 0.1) between soil NH_4^+ and NO production rate at $T_{soil}=10°C$, NO production rate at $T_{soil}=20°C$, and NO consumption coefficient at $T_{soil}=20°C$ was found. It was not possible to explain the differences in net potential NO fluxes through the physical and chemical soil parameters. However, we found a strong relationship between understory types and the amount of net potential NO fluxes. It may originate from the complex biological interactions between plants and their soil environment. Explanations for the differences in net potential NO fluxes could be:

- that roots can generate NO (Stöhr and Ullrich, 2002; Stöhr and Stremlau, 2006),
- that litter type and the influence of root exudates influenced functions of the soil microbial communities under the respective understory plants,
- that mycorrhiza influence the magnitude of net potential NO fluxes.

3.3 N₂O fluxes in the Hainich

We performed closed chamber measurements of N_2O over a period of one year (November 2005 to November 2006) in an unmanaged old growth beech forest (Hainich, Thuringia, Germany). To do justice to the spatial variability we measured N_2O fluxes at 15 plots distributed over the Hainich research site. During the measuring period we performed 510 gas flux measurements at 34 dates. These N_2O fluxes ranged between -101.7 and 121.6 µg N_2O m^{-2} h^{-1} (Fig. 8).

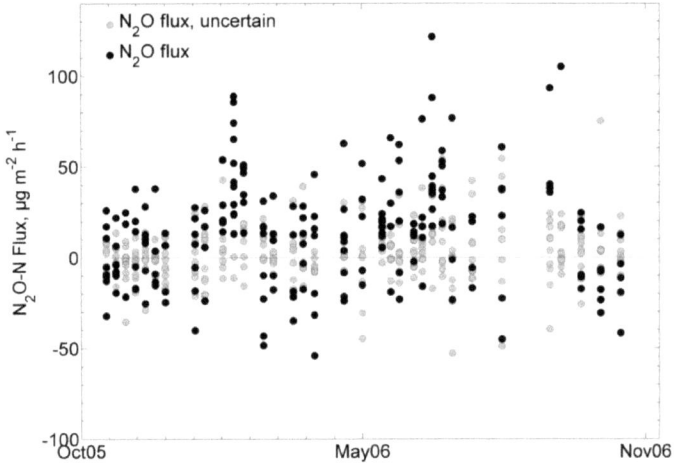

Figure 8: N_2O fluxes at the Hainich research site measured at 15 plots; black dots are the certain (r^2 of the linear regression of the concentration measurements at 0, 10 and 20 minutes after closure of the chamber not below 0.7) field N_2O fluxes and the grey dots display the uncertain N_2O fluxes. All N_2O fluxes are expressed in terms of mass of nitrogen.

The averaged field N_2O fluxes exhibited small amplitudes between -5.0 and 38.8 µg N_2O m^{-2} h^{-1}, but most average field N_2O fluxes did not significantly differ from zero (t-test, p = 0.05) (see Fig. 9). Through interpolating between the 34 days of measuring an annual N_2O flux was calculated. This annual N_2O flux was 0.46±0.32 kg N_2O ha^{-1} yr^{-1} for the Hainich research site. The highest field N_2O fluxes occurred between January and February 2006. During this time there was a frost period with soil temperature always below -0.5°C. This period contributes up to 40% to the annual field N_2O emission. However, the longest part of the year a background emission pattern was observed (mean N_2O flux during these periods: 3.4 µg N_2O m^{-2} h^{-1}).

Also, other studies performed on beech forest sites found mostly background emission patterns with annual N_2O emissions at or below 0.5 kg N_2O ha^{-1} yr^{-1} (Guckland, 2009; Brumme and Borken, 2009; Borken and Beese, 2006). Hence, a background flux around 0.5 kg N_2O ha^{-1} yr^{-1} for the given background emission type and the given background emission factor seems to be more adequate.

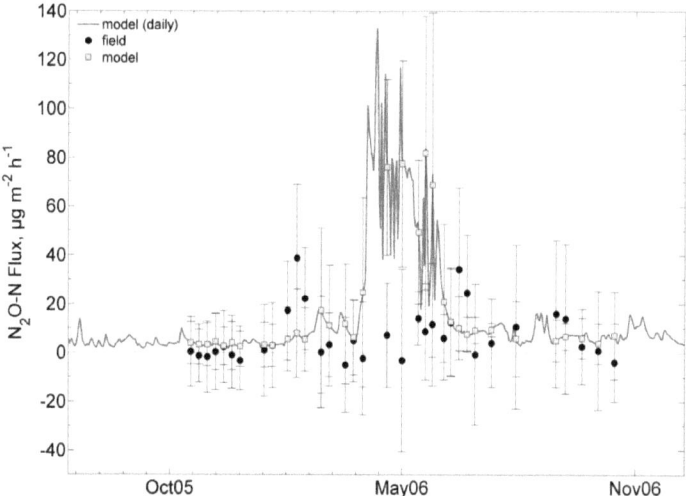

Figure 9: Field and modelled N_2O fluxes for the Hainich research site from November 2005 to November 2006. The black dots are the mean field N_2O flux rates (n=15), the grey squares are the mean modelled N_2O flux rates (n=15) and the grey line shows the mean daily modelled N_2O flux rates. The error bars on each individual data point are the standard deviation (n=15). All N_2O fluxes are expressed in terms of mass of nitrogen.

The spatial variability of the 15 plots at the Hainich research site was always high during the measuring period. There is a slightly positive correlation between the correlation of N_2O fluxes at two plots and the distances between these plots. However, these correlations are not significant.

A relationship between N_2O fluxes and physical and chemical soil parameters could not be observed. Only at six days (out of 34) of measuring period significant correlations between bulk density and N_2O fluxes could be found.

Synthesis

3.4 Modelling N_2O fluxes

The N_2O fluxes simulated with the Forest-DNDC model for the 15 plots showed fluxes between 0.0 and 255.7 µg N_2O m^{-2} h^{-1}. The model does not account for N_2O uptake. Mean modelled N_2O fluxes (derived from the mean of the modelled N_2O fluxes for each plot at the date of N_2O flux measurements, n=34) ranged between 2.5 and 81.7 µg N_2O m^{-2} h^{-1}. These modelled values are typically larger than the plot averaged N_2O fluxes of the field measurements (see Fig. 9). The mean daily modelled N_2O fluxes (1.4-133.1 µg N_2O m^{-2} h^{-1}) were up to three times larger than the field N_2O fluxes.

Mean modelled N_2O fluxes showed lowest values from November 2005 to March 2006 and from August to November 2006. Highest values occurred at the end of April 2006. Furthermore, the daily modelled N_2O fluxes showed a weak seasonal pattern.

The annual modelled N_2O emission at measurement intervals and the annual modelled N_2O emission in daily resolution for the Hainich research site for the one year measuring period (November 2005 to November 2006) were 1.77±1.82 and 1.56±0.006 kg N_2O ha^{-1}. This is 3-fold higher than the annual N_2O flux derived from measured values.

The agreement of the temporal variability of modelled and field N_2O fluxes is not good. That was also displayed by the model efficiency, which showed negative values for all 15 N_2O fluxes derived with the Forest-DNDC model. In general, the Forest-DNDC model overestimates the N_2O fluxes of the Hainich research site.

4 Conclusion

In this thesis results from NO and N_2O measurements at three different research sites are presented. The measurements were performed on soil samples in an automated laboratory system and in situ by a closed static chamber technique. The main findings of this thesis can be concluded as follows:

(1) It was shown by this work that the comparison between laboratory measurements of soil samples taken from the field site and field measurements using the MBR method brings good results. The NO fluxes measured with these different techniques are in good agreements. Also former studies showed good agreements between the laboratory measurements of NO and field measurements (e.g. Kirkman et al., 2002; van Dijk et al.,

2002). However, in these former studies chamber measurements in the field were used instead of the MBR method. The gain of the laboratory method is that it is a relatively low-priced and easily handled method to determine NO fluxes of a specific site. Also samples of different sites can be investigated parallel. That is half the battle to chamber or MBR measurements.

(2) The NO fluxes observed through the DMBR method showed no diurnal course in contrast to the laboratory measurements and were generally low. One reason for the lower NO fluxes observed by the DMBR method can be the chemical reactions between NO and O_3 during daytime (see eq. 1.1), affecting not only the layer between the two measuring levels but also the layer from top of the soil to the lowest measuring level. The laboratory NO flux corresponds to the surface of the soil, while the field NO flux is attributed to 0.55 m above ground. Because turbulence is very weak close to the ground, the residence time of NO in the layer between soil surface (0 m above ground) and 0.55 m above ground might be long (compared to the characteristic time scale of the reaction of NO with O_3). Therefore, a considerable vertical divergence of the NO flux can occur, which would reduce the NO flux with increasing distance from the soil surface to 0.55 m above ground. However, the NO_2 generated in equation 1.1 can through ultraviolet light convert back to NO and an oxygen radical (see eq. 1.2), which build together with an oxygen molecule O_3 (see eq. 1.3). For future experiments it would be useful to attribute the field NO flux to a height as low as possible above ground to minimize the time for chemical reactions during the vertical transport of the trace gases.

(3) Laboratory NO flux measurements of soil samples taken at the Weidenbrunnen site have shown the relevance of investigating the organic layer. This was the first European laboratory study on NO fluxes from the organic layer. Only Japanese scientist (Nishina et al., 2009) made a similar study one year before. The organic layer is in direct contact with the atmosphere and the majority of microorganisms live in the first centimeter of soil. Further studies on organic layers are necessary to understand the full influence on NO fluxes. For a better understanding it will be advisable to measure more soil and microbial parameters. The litter type and root exudates can influence the NO flux. Here, exact investigations on these points are important to make not only an educated guess. Also to measure more replicates is advisable. More replicates for example alleviated

Synthesis

better error estimation. Here we measured only two replicates as it is a very time consuming process.

(4) Net potential NO fluxes at the Weidenbrunnen site have shown a clear relationship with the different understory types at the site (moss, grass, young spruce and blueberries). Soil samples taken under moss and grass cover have shown small net potential NO fluxes while soil samples taken under spruce and blueberry cover have shown up to 10-fold higher net potential NO fluxes. Therefore, in further studies on biogenic NO emissions from forest floors more attention has to be paid to small scale heterogeneity of the understory vegetation.

(5) While the understory type seems to be an important parameter controlling NO exchange processes, corresponding soil nutrients played generally a less important role. It is remarkable that high NO emissions were observed for soils under woody understory types; this may be related to soil chemical processes in the vicinity of mycorrhized roots, but further studies are certainly necessary for confirmation. As forest thinning changes the availability of light and therefore the establishment of different understory types, management practices are likely to have important consequences on the net soil NO emission from a forested site.

(6) Results from the Hainich research site reveal particularly low N_2O emissions compared to the lowest values observed in managed beech forests. The results clearly underline that natural background emissions from this ecozone are lower than 1 kg N ha^{-1} yr^{-1}. Furthermore, the results indicate that site properties (soil pH, soil organic carbon, clay content, bulk density) substantially affect the magnitude of N_2O emissions. It is necessary in this respect to gain N_2O fluxes from more natural sites. Also, N_2O measurements with a higher time resolution would be favorable for a better understanding of the processes.

(7) The absence of spatial correlations indicates that within one beech site with nearly identical understory, the distance between each chamber is not as important. This is valuable for designing measurement plots as larger distances between individual chambers are not required. However, these findings need to be further verified by additional studies. A good way can be a study with a homogenized soil. In this way differences in the results based on soil properties can be excluded.

(8) This thesis also indicates that an uncalibrated Forest-DNDC model is not fully appropriate for simulating annual fluxes of N_2O for zero fertilizer treatments. To simulate the temporal variability in N_2O fluxes, a validation at other research sites seems to be necessary especially for sites with low soil organic carbon values. Therefore, it remains a challenge for future research to satisfactorily reproduce the spatial variability of natural N_2O fluxes. However, in this thesis the Forest-DNDC was used uncalibrated.

(9) NO and N_2O are intermediates in the nitrification and denitrification process. Both gases contribute to the greenhouse gas effect. It would have been good to measure NO and N_2O together at on site. Thereby statements of the interaction of both gases would have been possible.

5 References

Baldioli, M., Engel, T., Klöcking, B., Priesack, E., Schaaf, T., Sperr, C., and Wang, E.: Expert-N, ein Baukasten zur Simulation der Stickstoffdynamik in Boden und Pflanze, Lehreinheit für Ackerbau und Informatik im Pflanzenbau, TU München, Freisingen, 1994.

Bargsten, A., Falge, E., Huwe, B., and Meixner, F. X.: Laboratory measurements of nitric oxide release from forest soil with a thick organic layer under different understory types, Biogeosciences, 7, 1425-1441, 2010.

Baumgärtner, M., and Conrad, R.: Effects of soil variables and season on the production and consumption of nitric oxide in oxic soils, Biology and Fertility of Soils, 14, 166-174, 1992.

Behrendt, T.: A small-scale geostatistical analysis of the variability of soil properties driving the biogenic emission of nitric oxide from soil, MSc thesis, Geography, Johannes Gutenberg University Mainz, Mainz, Germany, 2009.

Beirle, S., Platt, U., Glasow, R., Wenig, M., and Wagner, T.: Estimates of nitrogen oxide emissions from shipping by satellite remote sensing, Geophysical Research Letters, 31, L18102, 10.1029/2004GL020312, 2004a.

Beirle, S., Platt, U., Wenig, M., and Wagner, T.: NO_x production by lightning estimated with GOME, Advances in Space Research, 34, 793-797, 2004b.

Beyrich, F., and Adam, W. K.: Site and Data Report for the Lindenberg Reference Site in CEOP – Phase I, Deutscher Wetterdienst, Offenbach230, 55, 2007.

Bollmann, A., and Conrad, R.: Influence of O_2 availability on NO and N_2O release by nitrification and denitrification in soils, Global Change Biology, 4, 387-396, 1998.

Borken, W., and Beese, F.: Methane and nitrous oxide fluxes of soils in pire and mixed stands of European beech and Norway spruce, European Journal of Soil Science, 57, 617-625, 10.1111/j.1365-2389.2005.00752.x, 2006.

Brady, N. C., and Weil, R. C.: The nature and properties of soils, 14 ed., Pearson Education, 2007.

Bremner, J. M., and Blackmer, A. M.: Effects of acetylene and soil water content on emission of nitrous oxide from soils, Nature, 280, 380-381, 1979.

Brumme, R., and Borken, W.: N_2O emission from temperate beech forest soils, in: Functioning and Management of European Beech Ecosystems, edited by: Brumme, R., and Khanna, P. K., Springer, Berlin, 353-367, 2009.

Butterbach-Bahl, K., Stange, F., and Papen, H.: Regional inventory of nitric oxide and nitrous oxide emissions for forest soils of southeast Germany using the biogeochemical model PnET-N-DNDC, Journal of Geophysical Research, 106, 34,155-134,166, 2001.

Chapuis-Lardy, L., Wrage, N., Metay, A., Chotte, J.-L., and Bernoux, M.: Soils, a sink for N_2O? A review, Global Change Biology, 13, 1-17, 2007.

Christensen, S.: Nitrous oxide emission from a soil under permanent grass: Seasonal and diurnal fluctuations as influenced by manuring and fertilization, Soil Biology and Biochemistry, 15, 531-536, 1983.

Conrad, R.: Soil microorganisms as controllers of atmospheric trace gases (H_2, CO, CH_4, OCS, N_2O, and NO), Microbiol. Rev., 60, 609-640, 1996.

Conrad, R.: Microbiological and biochemical background of production and consumption of NO and N2O in soil, in: Trace Gas Exchange in Forest Ecosystems, Tree Physiology (Series), Kluwer Academic Publ, Dordrecht, 3-33, 2002.

Crutzen, P. J.: Role of NO and NO_2 in the chemistry of the troposphere and stratosphere, Annual Review of Earth and Planetary Sciences, 7, 443-472, 1979.

Davidson, E. A., Keller, M., Erickson, H. E., Verchot, L. V., and Veldkamp, E.: Testing a conceptual model of soil emissions of nitrous and nitric oxides, BioScience, 50, 667-680, 2000.

Delon, C., Serca, D., Boissard, C., Dupont, R., Dutot, A., Laville, P., De Rosnay, P., and Delmas, R.: Soil NO emissions modelling using artificial neural network, Tellus B, 59, 502-513, 2007.

Denman, K. L., Brasseur, G. P., Chidthaisong, A., Ciais, P., Cox, P. M., Dickinson, R. E., Hauglustaine, D., Heinze, C., Holland, E. A., Jacob, D. J., Lohmann, U., Ramachandran, S., da Silva Dias, P. L., Wofsy, S. C., and Zhang, X.: Couplings between changes in the climate system and boigeochemistry, in: Climate Change 2007: The physical science basis. contribution of working group 1 to the fourth assesment report of the Intergovernmental Panel on Climate Change, edited by: Solomon, S., Qin, D., Manning, M., Chen, Z., Marquis, M., Averyt, K. B., Tignor, M., and Miller, H. L., University Press Cambridge, Cambridge, 2007.

Falge, E., Tenhunen, J. D., and Aubinet, M.: A model-based study of carbon fluxes at ten European forest sites, in: Fluxes of carbon, water and energy of European forests edited by: Valentini, R., Springer, Berlin, 151-177, 2003.

Feig, G. T., Mamtimin, B., and Meixner, F. X.: Soil biogenic emissions of nitric oxide from a semi-arid savanna in South Africa, Biogeosciences, 5, 1723-1738, 2008.

Flessa, H., Ruser, R., Schilling, R., Loftfield, N., Munch, J. C., Kaiser, E. A., and Beese, F.: N_2O and CH_4 fluxes in potato fields: automated measurement, management effects and temporal variation, Geoderma, 105, 307-325, 2002.

Foken, T.: Lufthygienisch-biologische Kennzeichnung des oberen Egertales (Fichtelbegirge bis Karlovy Vary), Bayreuther Forum Ökologie, 100, 1-70, 2003.

Galbally, I. E., and Johansson, C.: A model relating laboratory measurements of rates of nitric-oxide production and field-measurements of nitric-oxide emission from soils, Journal of Geophysical Research-Atmospheres, 94, 6473-6480, 1989.

Gerstenberger, P., Foken, T., and Kalbitz, K.: The Lehstenbach and Steinkreuz catchment in NE Bavaria, Germany, in: Biogeochemistry of forested catchments in a changing environment: a german case study, edited by: Matzner, E., Springer, Berlin, 15-44, 2004.

Guckland, A.: Nutrient stocks, acidity, processes of N transformation and net uptake of methane in soils of a temperate deciduous forest with dieffrent abundance of beech, PhD, Fakultät für Forstwissenschaften und Waldökologie, Georg-August-Universität Göttingen, Göttingen, 121 pp., 2009.

Johansson, C.: Field measurements of emission of nitric oxide from fertilized and unfertilized forest soils in Sweden, Journal of Atmospheric Chemistry, 1, 429-442, 1984.

Kiese, R., Li, C., Hilbert, D. W., Papen, H., and Butterbach-Bah, K.: Regional application of PnET-N-DNDC for estimating the N_2O source strength of tropical rainforests in the wet tropics of Australia, Global Change Biology, 11, 128-144, 10.1111/j.1365-2486.2004.00873.x, 2005.

Kirkman, G. A., Yang, W. X., and Meixner, F. X.: Biogenic nitric oxide emissions upscaling: an approach for Zimbabwe, Global Biogeochemical Cycle, 15, 2001.

Kirkman, G. A., Gut, A., Ammann, C., Gatti, L. V., Cordova, A. M., Moura, M. A. L., Andreae, M. O., and Meixner, F. X.: Surface exchange of nitric oxide, nitrogen dioxide, and ozone at a cattle pasture in Rondonia, Brazil, Journal of Geophysical Research-Atmospheres, 107, 10.1029/2001JD000523, 2002.

Li, C., Frolking, S. E., and Frolking, T. A.: A model of nitrous oxide evolution from soil driven by rainfall events: 1. model structure and sensitivity, Journal of Geophysical Research, 97, 9759-9776, 1992.

Li, C., Aber, J., Stange, F., Butterbach-Bahl, K., and Papen, H.: A process-oriented model of N_2O and NO emissions from forest soils: 1. Model development, Journal of Geophysical Research, 105, 2000.

Ludwig, J., and Meixner, F. X.: Surface exchange of nitric oxide (NO) over three European ecosystems, in: Proceedings of the sixth European symposium on the physico-chemical behaviour of atmospheric pollutants edited by: Angeletti, G., and Restelli, G., Commission of the European communities, Luxembourg, 587-593, 1994.

Ludwig, J., Meixner, F. X., Vogel, B., and Forstner, J.: Soil-air exchange of nitric oxide: An overview of processes, dnvironmental vactors, and modeling studies, Biogeochemistry, 52, 225-257, 2001.

Martin, R., and Asner, G.: Regional estimate of nitric oxide emissions following woody encroachment: linking imaging spectroscopy and field studies, Ecosystems, 8, 33-47, 2005.

Martin, R. V., Jacob, D. J., Chance, K., Kurosu, T. P., Palmer, P. I., and Mathew, J. E.: Global inventory of nitrogen oxide emissions constrained by space-based observations of NO_2 columns, Journal of Geophysical Research, 108, 4537, 10.1029/2003JD003453, 2003.

Mayer, J.-C., Bargsten, A., Rummel, U., Meixner, F. X., and Foken, T.: Distributed modified bowen ratio method for surface layer fluxes of reactive and non-reactive trace gases, Agricultural and Forest Meteorology, in press, 2010.

Mayer, J.-C., Bargsten, A., Rummel, U., Meixner, F. X., and Foken, T.: Distributed modified bowen ratio method for surface layer fluxes of reactive and non-reactive trace gases, Agricultural and Forest Meteorology, submitted.

Meixner, F. X., Fickinger, T., Marufu, L., Serça, D., Nathaus, F. J., Makina, E., Mukurumbira, L., and Andreae, M. O.: Preliminary results on nitric oxide emission from a southern African savanna ecosystem, Nutrient Cycling in Agroecosystems, 48, 123-138, 1997.

Meixner, F. X., and Yang, W.: Biogenic emissions of nitric oxide and nitrous oxide from arid and semi-arid land, in: Dryland Ecohydrology, edited by: D´Odoricoand, P., and Porporat, A., Springer, Dordrecht, 233-255, 2006.

Mosier, A. R., Parton, W. J., and Hutchinson, G. L.: Modelling nitrous oxide evolution from cropped and native soils, Ecological Bulletins, 229-241, 1983.

Nishina, K., Takenaka, C., and Ishizuka, S.: Spatial variations in nitrous oxide and nitric oxide emission potential on a slope of Japanese cedar (*cryptomeria japonica*) forest, Soil. Sci. Plant. Nutr., 55, 179–189, 2009.

Otter, L. B., Yang, W. X., Scholes, M. C., and Meixner, F. X.: Nitric oxide emissions from a southern African savanna, Journal of Geophysical Research-Atmospheres, 104, 18471-18485, 1999.

Parton, W. J., Mosier, A. R., Ojima, D. S., Valentine, D. W., Schimel, D. S., Weier, K., and Kulmala, A. E.: Generalized model for N_2 and N_2O production from nitrification and denitrification, Global Biogeochemical Cycles, 10, 401-412, 10.1029/96gb01455, 1996.

Pathak, H.: Emissions of nitrous oxide from soil, Current Science, 77, 359-369, 1999.

Pilegaard, K., Hummelshoj, P., and Jensen, N. O.: Nitric oxide emission from a Norway spruce forest floor, Journal of Geophysical Research-Atmospheres, 104, 3433-3445, 1999.

Potter, C. S., Klooster, S. A., and Chatfield, R. B.: Consumption and production of carbon monoxide in soils: A global model analysis of spatial and seasonal variation, Chemosphere, 33, 1175-1193, 1996.

Prather, M., and Ehhalt, D.: Atmospheric chemistry and greenhouse gases, Cambridge, UK, 239-287, 2001.

Remde, A., Ludwig, J., Meixner, F. X., and Conrad, R.: A study to explain the emission of nitric-oxide from a marsh soil, Journal of Atmospheric Chemistry, 17, 249-275, 1993.

Robertson, G. P., and Groffman, P. M.: Nitrogen transformation, in: Soil microbilogy, ecology, and biochemistry, edited by: Paul, E. A., Elsevier, Heidelberg, 2007.

Saggar, S., Andrew, R. M., Tate, K. R., Hedley, C. B., Rodda, N. J., and Townsend, J. A.: Modelling nitrous oxide emissions from dairy-grazed pastures, Nutrient Cycling in Agroecosystems, 68, 243-255, 2004.

Scheffer, F., and Schachtschabel, P.: Lehrbuch der Bodenkunde, Spektrum, Berlin, 593 pp., 2002.

Skiba, U., Fowler, D., and Smith, K.: Emissions of NO and N_2O from soils, Environmental Monitoring and Assessment, 31, 153-158, 1994.

Skiba, U., Fowler, D., and Smith, K. A.: Nitric oxide emissions from agricultural soils in temperate and tropical climates: sources, controls and mitigation options, Nutrient Cycling in Agroecosystems, 48, 139-153, 1997.

Skopp, J., Jawson, M. D., and Doran, J. W.: Steady-state aerobic microbial activity as a function of soil water content, Soil Science Society of America Journal, 54, 1619-1625, 1990.

Smith, K. A., Thomson, P. E., Clayton, H., McTaggart, I. P., and Conen, F.: Effects of temperature, water content and nitrogen fertilisation on emissions of nitrous oxide by soils, Atmospheric Environment, 32, 3301-3309, 1998.

Smith, K. A., Ball, T., Conen, F., Dobbie, K. E., Massheder, J., and Rey, A.: Exchange of greenhouse gases between soil and atmosphere: interactions of soil physical factors and biological processes, European Journal of Soil Science, 54, 779-791, 2003.

Steinkamp, J., Ganzeveld, L. N., Wilcke, W., and Lawrence, M. G.: Influence of modelled soil biogenic NO emissins on related trace gases and the atmospheric oxidizing efficiency, Atmospheric Chemistry and Physics, 9, 2663-2677, 2009.

Stöhr, C., and Ullrich, W. R.: Generation and possible roles of NO in plant roots and their apoplastic space, Journal of Experimental Botany, 53, 2293-2303, 2002.

Stöhr, C., and Stremlau, S.: Formation and possible roles of nitric oxide in plant roots, Journal of Experimental Botany, 57, 463-470, 2006.

Subke, J.-A., Reichstein, M., and Tenhunen, J. D.: Explaining temporal variation in soil CO_2 efflux in a mature spruce forest in Southern Germany, Soil Biology and Biochemistry, 35, 1467-1483, 2003.

van Dijk, S. M., Gut, A., Kirkman, G. A., Meixner, F. X., Andreae, M. O., and Gomes, B. M.: Biogenic NO emissions from forest and pasture soils: Relating laboratory studies to field measurements, Journal of Geophysical Research-Atmospheres, 107, 2002.

Williams, E. J., Hutchinson, G. L., and Fehsenfeld, F. C.: NO_x and N_2O emissions from soil, Global Biogeochemical Cycles, 6, 351-388, 10.1029/92gb02124, 1992.

Winkler, J. P., Cherry, R. S., and Schlesinger, W. H.: The Q_{10} relationship of microbial respiration in a temperate forest soil, Soil Biology and Biochemistry, 28, 1067-1072, 1996.

Yan, X., Ohara, T., and Akimoto, H.: Statistical modeling of global soil NO_x emissions, Global Biogeochemical Cycles, 19, GB3019, 2005.

Yang, W. X., and Meixner, F. X.: Laboratory studies on the release of nitric oxide from subtropical grassland soils: The effect of soil temperature and moisture, in: Gaseous nitrogen emissions from grasslands, edited by: Jarvis, S. C., and Pain, B. F., CAB International, Wallingford, UK, 67-71, 1997.

Yu, J. B., Meixner, F. X., Sun, W. D., Liang, Z. W., Chen, Y., Mamtimin, B., Wang, G. P., and Sun, Z. G.: Biogenic nitric oxide emission from saline sodic soils in a semiarid region, northeastern China: A laboratory study, Journal of Geophysical Research, 113, 11, 2008.

Appendix A

Individual Contributions to the Publications

The publications of which this cumulative thesis consists were composed in close cooperation with other researchers. Hence, other authors also contributed to the publications listed in appendices B to D in different ways. In this section I specify my own contributions to the individual manuscripts.

Appendix B

Authors	Bargsten, A.*, E. Falge, K. Pritsch, B. Huwe and F.X. Meixner
Title	Laboratory measurements of nitric oxide release from forest soil with a thick organic layer under different understory types
Status	published
Journal	Biogeoscience, 7, 1425-1441, 2010

Contribution

Bargsten	65%	idea, methods, data collection, data analysis, manuscript writing, figures, discussion, editing, corresponding author
Falge	15%	idea, discussion
Pritsch	5%	discussion
Huwe	5%	discussion
Meixner	10%	idea, discussion

Appendix A

Appendix C

Authors	Mayer, J.-C.*, A. Bargsten, U. Rummel, F.X. Meixner and T. Foken
Title	Distributed Modified Bowen Ratio Method for Surface Layer Fluxes of reactive and non-reactive Trace Gases
Status	submitted
Journal	Agricultural and Forest Meteorology

Contribution

Mayer	50%	idea, methods, data collection, data analysis, manuscript writing, figures, discussion, editing, corresponding author
Bargsten	35%	soil sample preparation, laboratory analysis, data analysis, manuscript writing
Rummel	5%	discussion
Meixner	5%	discussion
Foken	5%	idea, discussion

Appendix D

Authors	Bargsten, A.*, M. Timme, S. Glatzel, and H. Jungkunst
Title	Low nitrous oxide emissions in an unmanaged old growth beech forest
Status	in preparation
Journal	European Journal of Soil Science

Contribution

Bargsten	80%	idea, methods, data collection, data analysis, modeling, manuscript writing, figures, discussion, editing, corresponding author
Timme	5%	statistic analysis, discussion
Glatzel	5%	idea, discussion
Jungkunst	10%	idea, discussion

Appendix B

Laboratory measurements of nitric oxide release from forest soil with a thick organic layer under different understory types

A. Bargsten[1], E. Falge[1], K. Pritsch[2], B. Huwe[3] and F.X. Meixner[1,4]

[1]{Biogeochemistry Department, Max Planck Institute of Chemistry, 55020 Mainz, Germany

[2] {Institute of Soil Ecology, German Research Center for Environmental Health, Helmholtz Zentrum München, Neuherberg, Germany}

[3]{Soil Physics Department, University of Bayreuth, Germany}

[4]{Physics Department, University of Zimbabwe, Harare, Zimbabwe}

Correspondence to: A. Bargsten ()

Submitted to Biogeoscience: December 2009

Published in Biogeoscience Discussions: January 2010

Appendix B

Abstract

Nitric oxide (NO) plays an important role in the photochemistry of the troposphere. NO from soil contributes up to 40% to the global budget of atmospheric NO. Soil NO emissions are primarily caused by biological activity (nitrification and denitrification), that occurs in the uppermost centimeter of the soil, a soil region often characterized by high contents of organic material. Most studies of NO emission potentials to date have investigated mineral soil layers. In our study we sampled soil organic matter under different understories (moss, grass, spruce and blueberries) in a humid mountainous Norway spruce forest plantation in the Fichtelgebirge (Germany). We performed laboratory incubation and flushing experiments using a customized chamber technique to determine the response of net potential NO flux to physical and chemical soil conditions (water content and temperature, bulk density, particle density, pH, C/N ratio, organic C, soil ammonium, soil nitrate). Net potential NO fluxes (in terms of mass of N) from soil samples taken under different understories ranged from 1.7-9.8 ng m^{-2} s^{-1} (soil sampled under grass and moss cover), 55.4-59.3 ng m^{-2} s^{-1} (soil sampled under spruce cover), and 43.7-114.6 ng m^{-2} s^{-1} (soil sampled under blueberry cover) at optimum water content and a soil temperature of 10°C. The water content for optimum net potential NO flux ranged between 0.76 and 0.8 gravimetric soil moisture for moss covered soils, between 1.0 and 1.1 for grass covered soils, 1.1 and 1.2 for spruce covered soils, and 1.3 and 1.9 for blueberry covered soils. Effects of soil physical and chemical characteristics on net potential NO flux were statistically significant (0.01 probability level) only for NH_4^+. Therefore, as an alternative explanation for the differences in soil biogenic NO emission we consider more biological factors like understory vegetation type, amount of roots, and degree of mycorrhization; they have the potential to explain the observed differences of net potential NO fluxes.

1 Introduction

Nitric oxide (NO) is a reactive gas which plays a central role in the photochemistry of the troposphere (Crutzen, 1979). The photochemistry of NO and nitrogen dioxide (NO_2) is important for the generation/destruction of tropospheric ozone and, hence, regulates the oxidizing capacity of the troposphere. The oxidation products of NO (gaseous NO_2, nitrous and nitric acid, particulate nitrite and nitrate) also contribute to the generation of acid rain (Crutzen, 1979) affecting human health and plant productivity.

With respect to NO biosphere-atmosphere exchange, soils are of great interest due to the fact that NO biogenic emissions from soil contribute up to 40% to the global budget of atmospheric NO (Davidson and Kingerlee, 1997; Meixner, 1994; Denman et al., 2007; Rudolph and Conrad, 1996). **Kesik et al. (2005) predicted that by 2039 soil NO emissions will increase by 9%.** Soils have the potential for acting as a sink for atmospheric NO (Conrad, 1994). Only a few studies provide an indication of soils acting as a sink (Dunfield and Knowles, 1998; Skiba et al., 1994; Slemr and Seiler, 1991). The NO flux between soil and atmosphere is a result of microbial consumption and production of NO in the top soil layer. NO production and consumption occur simultaneously during nitrification and denitrification (Remde et al., 1989; Rudolph and Conrad, 1996; Skiba et al., 1997; Firestone and Davidson, 1989). In both soil microbial processes NO can be an intermediate, it can be released and also absorbed (Galbally, 1989).

In most cases the organic layer is the only soil layer in direct contact with the atmosphere. There are soils having an organic layer with a thickness of 10 cm or more; these thick organic layers are mostly a kind of moder or raw humus (Scheffer and Schachtschabel, 2002). Mineral soils under these organic layers are never in contact with the atmosphere. Hence, as shown by Gasche and Papen (1999), who examined soils under a spruce canopy, the most important layer for NO exchange is the uppermost organic layer. In their experiment with intact soil cores from a spruce forest site they found that the organic layer contributed over 86% to the NO emission from soil. It is also known that nitrification occurs predominantly in the first few centimeters of soils (Papke and Papen, 1998; Rudolph and Conrad, 1996; Laville et al., 2009; Venterea et al., 2005; Remde et al., 1993; Jambert et al., 1994). Venterea et al. (2005) found actually the highest NO production in the first centimeter. Organic soils support high nitrification and denitrification rates and may be important hot spots of NO emission (Guthrie and Duxbury, 1978). Denitrification, in contrast, normally occurs in deeper soil layers or in the water table. In this respect, the role of organic matter is potentially important (Jambert et al., 1994).

In forests the type of understory influences NO exchange between the soil and the trunk space (Jambert et al., 1994; Pilegaard et al., 1999). Most studies to date have focused on the influence of the overstory vegetation and/or soil nutrients (Fowler et al., 2009; Venterea et al., 2004; Pilegaard et al., 2006). As reported by Oberdorfer (1994), Norway spruce forests fall into a series of plant sociological associations, which are characterized by the main understory species present (e.g. *Calamagrostio villosae - Piceetum*). Within one individual forest stand the understory might be composed of patches characterized by different species (e.g. *Calamagrostis villosa, Vaccinium myrtillus, Deschampsia flexuosa*). There are only a few studies how plants influence the NO exchange between soil and atmosphere (Stöhr and Stremlau, 2006; Stöhr and Ullrich, 2002), and there is a considerable lack of knowledge in this area.

To investigate the effect of soil physical and chemical parameters and understory types on NO emission from thick organic layers of forest soils we carried out laboratory incubation and flushing experiments on soils sampled below various understory covers in a Norway spruce forest in south-eastern Germany.

2 Material and methods

2.1 Sample Site

The field site is located at Weidenbrunnen (50°09' N, 11°34' E, 774 m above sea level) which is situated in the Fichtelgebirge Mountains, NE Bavaria, Germany. The site is mainly covered by 55-year-old Norway spruce (*Picea abies*) with significant variability in the understory. There are four different main understory types: moss, grass (*Deschampsia flexuosa* and *Calamagrostis villosa*), blueberries (*Vaccinium myrtillus*), and young spruce which cover 45, 19, 7 and 13%, respectively, of the total surface area of the Weidenbrunnen site (Behrendt,

2009). Mean annual air temperature of the Weidenbrunnen site is 5.3°C, mean annual soil temperature is 6.3°C, and mean annual precipitation is approximately 1160 mm (1971 - 2000; Foken, 2003; Falge et al., 2003). The soil type was classified as cambic podzol over granite (Subke et al., 2003), and the texture is sandy loam to loam, with relatively high clay content in the Bh horizon. The mineral soil is characterised by low pH values (<4). The soil litter and the organic horizon had a thickness between 5 and 9 cm (Behrendt, 2009). The organic layer is classified as a moder consisting of Oi, Oe, and Oa horizons. More details concerning the site can be found in Gerstberger et al. (2004).

2.2 Soil sampling and preparation

In September 2008, soil samples for the laboratory study on NO release were taken from the O horizon at patches below the main understory types: moss, grass, young spruce, and blueberries. An individual understory patch has been defined, such that one square meter of understory area has to be covered mainly (>50%) with the respective understory vegetation. Two samples were taken for each understory type, resulting in a total of eight soil samples (soil samples taken under moss: M1, M2, soil samples taken under grass: G1, G2, soil samples taken under spruce: S1, S2, soil samples taken under blueberries: B1, B2). The soil samples were air dried and then stored at 4°C until analysis. All measurements were performed within 2 months after sampling.

For our laboratory studies of NO release rates, samples were sieved through a 16 mm mesh to homogenise the soil and, all green biomass was removed. This can be contrasted with previous studies of mineral soils and sands where samples were sieved through 2 mm mesh (van Dijk and Meixner, 2001; van Dijk et al., 2002; Feig et al., 2008; Yu et al., 2008; Gelfand et al., 2009). A 16 mm mesh was chosen, based on tests sieving Weidenbrunnen organic matter through 2, 4, 8, and 16 mm mesh sizes. These experiments showed, that sieving through a 2 mm mesh destroyed the structure of soil organic matter causing higher NO release rates than observed when sieving through 4, 8 and 16 mm meshes whose corresponding NO release rates were not significantly different from each other (see Fig. 1).

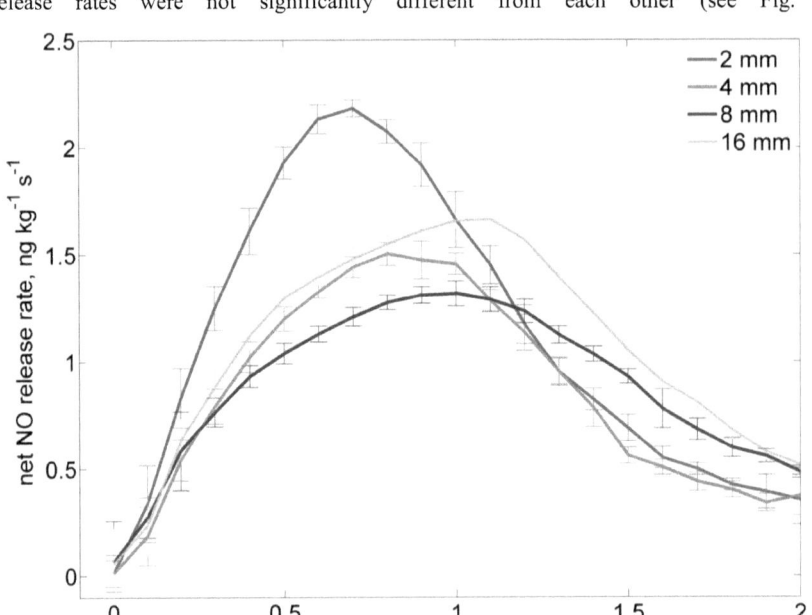

Figure 1. The effect of sieving organic soil samples through sieves of different mesh sizes on the observed net NO release rates ($T_{soil}=10°C$). Error bars show the standard deviation of the net NO release rate (expressed in terms of mass of nitrogen) averaged over bins of 0.1 gravimetric soil moisture.

For measurements of net NO release rates approx. 0.1 kg soil was placed into a Plexiglas cuvette, wetted with deionised water to a gravimetric water content >3 (using a spray can) and pre-incubated for 3 hours in a thermo-regulated cabinet to adapt to the soil temperature used during the corresponding NO release experiments. Former experiments showed that net NO release rates increase fairly proportionally with soil mass in the chambers up to 100 g, after which the slope declines. This indicates that from this soil mass onwards gas diffusion through the soil could be limiting. These results are similar to those of Remde et al. (1989)

where the NO flux rate was shown to be proportional to the soil mass in the chamber up to 150 g. Above 150 g the relationship between NO flux and soil mass was no longer linear.

2.3 Soil physical and chemical characterization

In addition to samples for use in flux measurements, we took organic layer samples from each understory patch for the determination of soil pH, C/N ratio, organic C (C_{org}), soil nitrate (NO_3^-), soil ammonium (NH_4^+), bulk density (BD) and particle density (PD).

For the determination of soil pH the organic matter was homogenized and afterwards measured in a soil-to-water suspension (1:2.5) using a glass electrode (SenTix®, WTW, Germany). The C/N ratio was measured with an elementary analyzer (Flash EA 1112, Thermoquest, Germany). C_{org} was determined by the mean difference of 5 g (air dried) of the soil sample and 5 g dried at 430°C in a muffle furnace (until constant weight was achieved). The ammonium and nitrate concentrations in extracts of the soil samples were measured by spectrometry (FIA-lab, MLE, Germany). For determination of the soil bulk density, undisturbed soil samples were taken using a spade and afterwards dimensioned. Then the samples were dried at 60°C for 24 h. From each patch we took three soil cores and individual quantities were averaged over these. Particle density of the soil sample was determined by a heliumpycnometer (AccuPyc II 1340, Micromeritics, USA) after sieving soil samples through a 2 mm mesh.

2.4 Laboratory setup

Net NO release rates from soil samples were determined using an automated laboratory system. A detailed description of our experimental setup is given in van Dijk and Meixner (2001); here we give only a short description of the most recent state of the setup (see Fig. 2).

Pressurized air is passed through a pure air generator (PAG 003, ECOPHYSICS, Switzerland) to provide dry and NO-free air. This NO-free air supplied five Plexiglas cuvettes (four incubation cuvettes and one empty reference cuvette). The volume of each cuvette was $9.7*10^{-4}$ m^3 (0.97 l)) and each was flushed with a continuous flow of $4.2*10^{-5}$ m^3 s^{-1} (2.5 l min^{-1}) of dry NO-free air, as controlled by five mass flow controllers (MFC, Mass-Flo®, 5000 sccm range, MKS instruments, USA), one for each cuvette. The headspace volume of each cuvette is well mixed by a teflonized micro-fan (Micronel®, USA). The outlet of each cuvette was connected to a switching valve. Every two minutes one cuvette was switched to be the "active" cuvette (i.e., connected to the analyzers, while the remaining four cuvettes were still purged), so that all five cuvettes were measured within 10 minutes. The valves provided necessary sample air to a chemiluminescence detector, NO-analyser (Model 42i Trace Level, Thermo Electron Corporation, USA; detection limit: 250 ppt (3σ)) and a CO_2-/H_2O-analyzer (Li-cor 840, Licor, USA). Instead of ambient air we operated the NO-analyser with pure oxygen (O_2) to obtain a better accuracy and precision of the NO mixing ratio measurements, particularly at low mixing ratios.

The NO-analyser was calibrated using a gas phase titration unit (GPT, 146 C Dynamic Gas Calibrator, Thermo Electron Corporation, USA). For operating the GPT we used NO-free air from the PAG 003 and an NO gas standard (5.02 ppm NO, Air Liquide, Germany). The determination of the soil NO compensation mixing ratio (Conrad, 1994) requires the flushing of incubated soil samples with enhanced NO mixing ratios (resulting in reduced or even negative net NO release rates, i.e. NO uptake by the soil). Hence, NO standard gas (200 ppm NO, Air Liquide, Germany) was diluted into the air flow from the PAG 003 via a mass flow controller (Flow EL, Bronkhorst, Germany).

All connections and tubes consisted of polytetrafluorethylene (PTFE). A homebuilt control unit (V25) was controlling the entire laboratory system and, in combination with a computer, was also used for data acquisition (see Fig. 2).

To determine the temperature response of the net NO release we performed a total of four experiments, each on another sub-sample of the original understory soil sample. The sub-samples were identically pre-treated. Incubations were at 10°C and 20°C, corresponding flushing was either with dry, NO-free air, or with air containing 133 ppb of NO. Since every experiment begins with a wetted soil sample and the flushing air is completely dry, the gravimetric water content (θ) of the samples declines during each experiment as evaporating water leaves the cuvette with the flushing air flow. Gravimetric soil moisture content was measured by tracking the loss of water vapour throughout the measurement period and relating this temporal integral to the gravimetric soil moisture content observed at the start and end of the measurement period. Soil samples are completely dry within 4 to 7 days. This procedure provides us the response of the net NO release rates over the entire range of gravimetric soil moisture (>4 to 0). Gravimetric soil moisture ranging from 0 to 4 corresponds to a water filled pore space (WFPS) from 0 to 0.7.

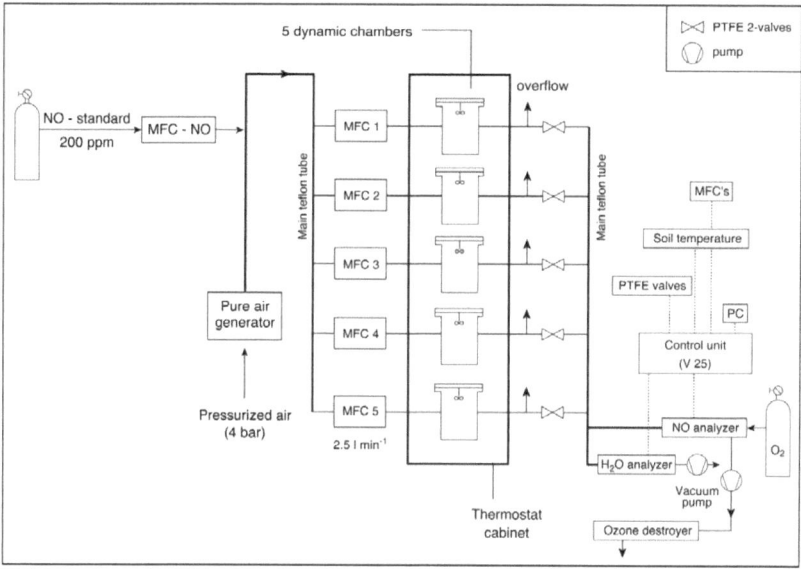

Figure 2. Experimental setup for laboratory investigation of net NO release rates on soil samples (details, see section 2.4).

The NO release rate is a product of NO consumption and NO production, because both processes occur simultaneously in the topsoil (Rudolph and Conrad, 1996; Conrad, 1994). Consequently, the observed NO release rate, J (see Eq. 1), is always a net release rate. If NO consumption overrides the NO production in the soil sample, then J becomes negative. However, this only occurs if the the NO mixing ratio in the reference cuvette, $m_{NO,ref}$, exceeds the NO mixing ratio in the headspace of a sample cuvette (which is equal to the corresponding outlet NO mixing ratio, $m_{NO,out}$, due to well-mixed conditions within each sample cuvette).

Appendix B

2.5 Calculation and fitting the net NO release rate

For a given constant incubation temperature (10°C, 20°C) we derived from our laboratory data the net NO release rate $J=J(\theta)$ (in ng NO (in terms of mass of nitrogen) per mass of (dry) soil (kg) and time (s)) as a function of the gravimetric soil moisture (θ) of the soil samples. $J(\theta)$ was calculated from the NO mixing ratio difference between the reference cuvette ($m_{NO,ref}$, in ppb) and the soil incubation cuvettes ($m_{NO,out}$, in ppb):

$$J(\theta) = \frac{Q}{M_{soil}}(m_{NO,out} - m_{NO,ref}) \times \frac{M_N}{V_m} \times 10^{-3} \quad (1)$$

where Q is the flow through the cuvette ($m^3\ s^{-1}$), M_{soil} is the dry mass of the soil sample (kg), $M_N/V_m * 10^{-3}$ is the conversion factor (ppb to ng m^{-3}), where M_N is the molecular weight of nitrogen (14.0076 kg $kmol^{-1}$) and V_m is the molar volume ($m^3\ kmol^{-1}$) at actual temperature and standard pressure (1013.25 hPa).

Individual data of measured net NO release rates were fitted with a 3 parameter function (Eq. 2) modified from that given by Meixner and Yang (2006) in order to yield two of the three parameters as measured quantities (θ_{opt}, J_{opt}):

$$J(\theta) = J_{opt} \times (\frac{\theta}{\theta_{opt}})^b \times \exp(b\left(1 - \frac{\theta}{\theta_{opt}}\right)) \quad (2)$$

where θ_{opt} is the gravimetric water content where the optimum net NO release rate (J_{opt} :=$J(\theta_{opt})$) is observed, and b characterizes the width of the fitting curve. The gnuplot® software (www.gnuplot.info, see copyright information) was used for fitting.

It has been frequently shown, that there is a linear relationship between the net NO release rate (J) and the headspace NO mixing ratio ($m_{NO,out}$) (Remde et al., 1989; van Dijk and Meixner, 2001; van Dijk et al., 2002; Ludwig et al., 2001):

$$J = P - K = P - k \times m_{NO,out} \times \frac{M_N}{V_m} \times 10^{-3} \quad (3)$$

Eq. (3) implies that the NO production rate P (ng $kg^{-1}\ s^{-1}$) is independent of the cuvette's headspace NO mixing ratio ($m_{NO,out}$), whereas the first-order NO consumption rate, K (ng $kg^{-1}\ s^{-1}$), is dependent on it. The NO consumption coefficient k ($m^3\ kg^{-1}\ s^{-1}$) is determined from the slope of Eq. (3). To obtain this slope, we used two incubation data sets: namely at $m_{NO,ref}$=0 ppb and $m_{NO,ref}$=133 ppb,

$$k(\theta) = \frac{\Delta J_{NO}}{\Delta[NO]} = \frac{J(m_{NO,out,high}) - J(m_{NO,out,low})}{m_{NO,out,high} - m_{NO,out,low}} \times \frac{V_m}{M_N} \times 10^{-3} \quad (4)$$

where $m_{NO,out,low}$ is the actual NO mixing ratio (ppb) in the headspace of the cuvette under flushing with NO free air and $m_{NO,out,high}$ is the actual NO mixing ratio in the cuvette under flushing with 133 ppb NO. Having determined k, the NO production rate P was calculated from Eq. (3) and corresponding NO net release rates J from Eq. (1).

Finally, Eq. (3) is extended to describe the net NO release rate, for each soil sample, as a function of the main influencing variables, headspace NO mixing ratio ($m_{NO,out}$), gravimetric water content (θ) and soil temperature (T_{soil}). For the temperature dependence we used the Q_{10} values (see section 2.8), as a "temperature amplification factor" (Feig et al., 2008):

$$J(m_{NO,out}, \theta, T_{soil}) = P(\theta, T_{soil}) - k(\theta, T_{soil}) \times m_{NO,out} \times \frac{M_N}{V_m} \times 10^{-3} \quad (5)$$

2.6 NO compensation point mixing ratio

The existence of a NO compensation point mixing ratio ($m_{NO,comp}$) has been clearly demonstrated (Remde et al., 1989; van Dijk and Meixner, 2001; Conrad, 1994; Gelfand et al., 2009; Feig et al., 2008; Otter et al., 1999; Johansson and Granat, 1984). Considering Eq. (5) $m_{NO,comp}$ is the mixing ratio ($m_{NO,out}$) at which the rate of NO production P equals the rate of NO consumption K, so that the net NO release rate between soil and the headspace is zero (J=0). Hence, from Eq. (6) $m_{NO,comp}$ is calculated in terms of gravimetric soil water content and soil temperature.

$$m_{NO,comp}(\theta, T_{soil}) = \frac{P(\theta, T_{soil})}{k(\theta, T_{soil})} \times \frac{V_m}{M_N} \times 10^{-3} \quad (6)$$

2.7 Net potential NO flux

To relate the net NO release rate, which is expressed in ng NO per mass of soil and time, to the net potential NO flux, which is expressed in ng NO per soil area and time, we used the following equation, originally presented by Galbally and Johansson (1989), which has been used in modified forms already by Otter et al. (1999), van Dijk and Meixner (2001), Feig et al. (2008), Gelfand (2009), Yu et al. (2008).

Appendix B

$$F_{NO}(\theta, T_{soil}) = \sqrt{D_p(\theta) \times BD \times k(\theta, T_{soil})} \times (\frac{P(\theta, T_{soil})}{k(\theta, T_{soil})} - m_{NO,out} \times \frac{M_N}{V_m} \times 10^{-3}) \qquad (7)$$

F_{NO} is the desired net potential NO flux (ng m^{-2} s^{-1}), BD is the bulk density of soil (kg m^{-3}), D_p is the effective diffusion coefficient of NO in soil (in m^2 s^{-1}) according to Millington and Quirk (1960) (see section 2.9).

2.8 Calculation of the Q_{10} value

The temperature dependence of the net potential NO flux was determined by using net NO release rates obtained for two soil temperatures, namely those at 10°C and 20°C. The temperature dependence usually shows an exponential increase and can be expressed by the ratio of two net potential NO fluxes, at soil temperatures 10°C apart. The Q_{10} values used for this study were calculated from the net potential NO fluxes at optimum gravimetric soil moisture (θ_{opt}):

$$Q_{10}(\theta_{opt}) = \frac{F_{NO}(\theta_{opt}, T_{soil}=20°C)}{F_{NO}(\theta_{opt}, T_{soil}=10°C)} \qquad (8)$$

2.9 Effective diffusion of NO in soil air

The effective gas diffusion coefficient of NO in soil air is an important parameter for deriving the net potential NO flux from NO production and NO consumption rates (Bollmann and Conrad, 1998). Since we do not have measurements of the effective soil diffusion coefficient (D_p) at the Weidenbrunnen site, we estimated the diffusion coefficient through available functional relationships. The choice of the proper diffusivity coefficient function is not trivial, particularly for organic soils (Kapiluto et al., 2007). Therefore, we tested different functions namely those of Moldrup et al. (2000), Millington (1959) and Millington and Quirk (1960) which are given in Tab. 1. In these functions the following measured variables were used:

- soil total porosity (Φ), calculated from the soil bulk density (BD) and the particle density (PD) of the soil sample; both parameters measured directly on the soil samples:

$$\Phi = 1 - \frac{BD}{PD} \qquad (9)$$

Appendix B

- soil air filled porosity (ε) calculated from the soil bulk density, the density of water (WD), and the the soil total porosity (Φ):

$$\varepsilon = 1 - \theta \frac{BD}{WD} \times \frac{1}{\Phi} \qquad (10)$$

We calculated net potential NO fluxes (see section 2.7) using the three different effective NO diffusion coefficients. One example for a soil sample from a grass covered patch is shown in Fig. 3. Net potential fluxes exhibit different maxima with a shifting value for the optimum water content for NO production due to the different exponents for ε. According to Moldrup (personal communication, 2009), the Millington and Quirk approach describes the effective gas diffusion coefficient best for soil organic matter; therefore the potential NO fluxes of this paper have been calculated using the formulation by Millington and Quirk (1960).

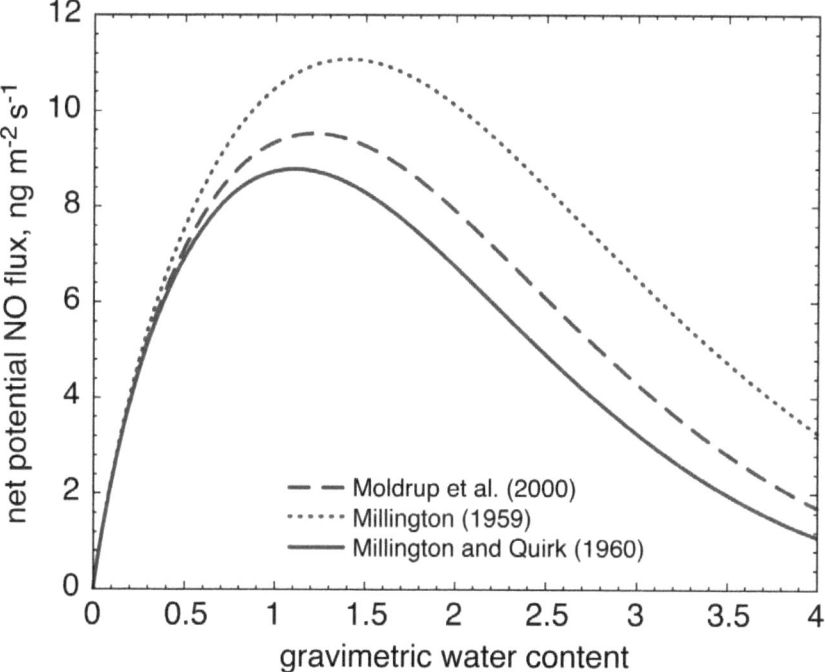

Figure 3. Net potential NO flux at 10°C from a grass covered patch (all expressed in terms of mass of nitrogen). The net potential NO fluxes were calculated according to Eq. (8) applying

Appendix B

effective soil diffusion coefficients by Moldrup et al. (2000), Millington (1959) and Millington and Quirk (1960) (see Tab. 1).

Table 1. Mathematical formulations for the calculation of the effective diffusion coefficient in soil. ϵ is the soil air-filled porosity in m^3 (soil air) m^{-3} (soil), Φ is the soil total porosity in m^3 (pores) m^{-3} (soil) and D$_0$ is the gas diffusion coefficient in free air (1.99*10^{-5} m^2 s^{-1}).

Moldrup (2000)	Millington (1959)	Millington & Quirk (1961)
$D_p = \dfrac{\epsilon^{2.5}}{\Phi}$	$D_p = \epsilon^{3/2} \times D_0$	$D_p = \dfrac{\epsilon^{10/3}}{\Phi^2} \times D_0$

2.10 Error estimation of NO release measurements

The errors in the net NO release rate were determined using the individual errors of all quantities on the right hand site of Eq. (1). We specified these errors as followed:

- The error in the soil weight (M$_{soil}$) measurements was set to the accuracy of the balance (PG-S Delta Range®, Mettler-Toledo, Switzerland) provided by the manufacturer: 0.001 kg (for a mass <1.0 kg).
- The error of the mass flow rate through the cuvette (Q) was found as 1.68*10^{-8} m^3 s^{-1} (i.e. the standard deviation of all individual mean flux rates of a corresponding experiment with n=798).
- The error of the mixing ratio in the headspace of a soil cuvette (m$_{NO,out}$) was determined by using every NO mixing ratio measurement: for m$_{NO,ref}$=0 ppb the error was <0.1 ppb, for m$_{NO,ref}$=133 ppb the error was <0.6 ppb.

- The error of the reference cuvette ($m_{NO,ref}$) was determined in the same way, resulting in an error of <0.1 ppb ($m_{NO,ref}$=0 ppb), and <0.4 ppb ($m_{NO,ref}$=133 ppb).

Application of Gaussian error propagation to Eq. (1) resulted in an error in the optimum net NO release rate (J_{opt}) of less than 8%.

The detection limit for the net NO release rate obtained by our laboratory system was determined by Feig et al. (2008) and Gelfand et al. (2009) using inert glass beads and autoclaved soils. The "blank" net NO release rate from the inert glass beads was 0.02 ng kg^{-1} s^{-1} with a random deviation of 0.02 ng kg^{-1} s^{-1} and for autoclaved soils it was 0.05 ng kg^{-1} s^{-1} with a random deviation of 0.02 ng kg^{-1} s^{-1}. Feig et al. (2008) defined the detection limit of the net NO release rate as 0.08 ng kg^{-1} s^{-1} (i.e. mean net NO release rate of glass beads plus three times its standard deviation). The detection limit of the autoclaved soils was calculated the same way and resulted in a detection limit of 0.11 ng kg^{-1} s^{-1}. Therefore, the more conservative estimate from the autoclaved soils was used as the detection limit of net NO release rates determined by our laboratory system.

In Fig. 4a and b, we present the net NO release rate calculated from the difference in the data points of NO mixing ratio (see Eq. 1) and the corresponding fit (see Eq. 2) for a soil samples under moss. Fig. 4 also shows the individual errors of J (by Gaussian error propagation; grey whiskers) and the detection limit of J (grey shadow band).

For the fit of the data according to Eq. (2), prediction bands (PB) were calculated at a confidence level of 95% using the procedure given by Olive (2007) (Eq. 2.6 in the work by Olive). The prediction bands show for a prescribed probability, the values of one or more hypothetical observations that could be drawn from the same population from which the given data was sampled.

Appendix B

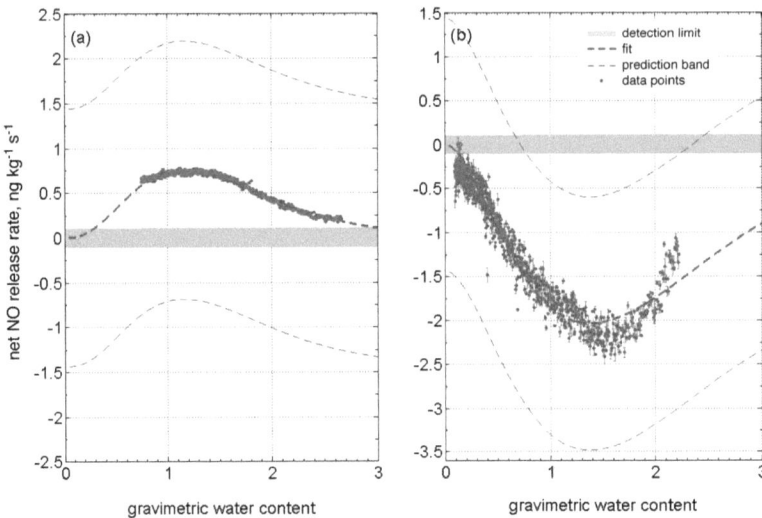

Figure 4. (a) measured net NO release rates (red dots) at $m_{NO,ref}=0$ ppb and fitted net NO release rates (red dashed line), for a soil samples covered with moss. (b) measured net NO release rates (red dots) at $m_{NO,ref}=133$ ppb and fitted net NO release rates (red dashed line), for a moss covered soil. The grey shaded band indicates the detection of the net NO release rate obtained through our laboratory system. Error bars (grey whiskers) on each individual data point have been calculated by the Gaussian error propagation (see section 2.10). NO release rates in both panels have been obtained for $T_{soil}=20°C$.

3 Results

3.1 Net NO release rates

Figures 5a-d present net NO release rates obtained from soil samples taken under moss, grass, spruce and blueberry cover at two temperatures (upper panels: 10°C, and lower panels: 20°C) and two NO mixing ratios (left panels: $m_{NO,ref}=0$ ppb and right panels: $m_{NO,ref}=133$ ppb). The curves are the result of corresponding fitting (Eq. 2) to measured data as described in section 2.5. At incubation with NO free air higher net NO release rates occurred from soil samples taken under spruce (S1, S2) and blueberry (B1, B2) cover than under moss (M1, M2) and grass (G1, G2) cover. Maximum NO release rates at $T_{soil}=10°C$ and 20°C were 12.4 and 23.6 ng kg^{-1} s^{-1} for S1, 13.2 and 32.0 ng kg^{-1} s^{-1} for S2, 11.4 and 25.5 ng kg^{-1} s^{-1} for B1, and 14.6 and 33.6 ng kg^{-1} s^{-1} for B2. Similarly, when incubated with 133 ppb NO, soil samples taken under spruce (S1: 9.3 and 14.4 ng kg^{-1} s^{-1}, S2: 10.4 and 30.8 ng kg^{-1} s^{-1}) and blueberry (B1: 6.8 and 23.6 ng kg^{-1} s^{-1}, B2: 13.6 and 30.2 ng kg^{-1} s^{-1}) cover showed the highest net NO release rates. In contrast, soil samples taken under moss and grass cover showed small net NO release rates when flushed with NO free air (Fig. 5a, c). When flushed with air containing 133 ppb NO, negative net NO release rates occurred for the soil samples S1, S2 and G2. In these cases the flushing NO mixing ratio of 133 ppb was obviously higher than the NO compensation mixing ratio ($m_{NO,comp}$) of the corresponding soil samples (see section 2.7), and the NO consumption rate (K) has exceeded the NO production rate (P) in these soil samples.

Net NO release rates reached their maxima between 0.64 (G1) and 2.41 (B2) gravimetric water content. The soil moisture, where the optimum net NO release rate is observed, is called the optimum soil moisture (θ_{opt} in Eq. 3). Generally, highest values of θ_{opt} were observed for S1, S2, B1 and B2.

At gravimetric soil moisture of 4 the net NO release rates do not become zero. That is due to the fact that the samples were not waterlogged at gravimetric soil moisture of 4. Therefore, nitrifiers and denitrifiers might be still supplied with oxygen.

However, the curves differ for optimum soil moistures and higher than these. The net NO release rates from S1 and S2 were not significantly different from each other using either flushing at $T_{soil}=10°C$, but significantly differ at $T_{soil}=20°C$. No significant differences could be observed between the two samples taken under moss cover, or the two samples taken under grass cover. Net NO release rate of soil samples taken under blueberry cover were similar only in a range between 0 and 1.4 gravimetric water content and only in the treatment with NO free air and at $T_{soil}=10°C$.

Appendix B

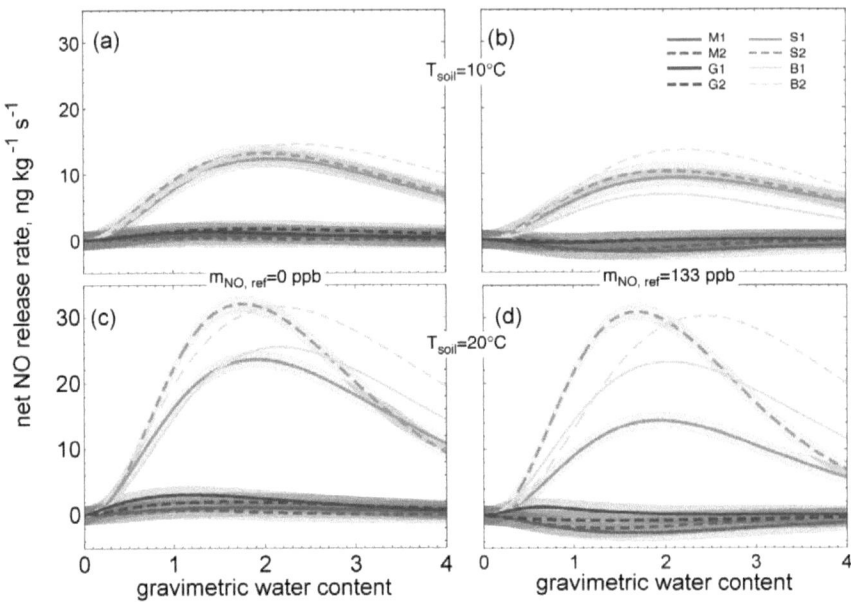

Figure 5. Net NO release rates fitted through experimental results by Eq. (2) (see section 2.4) at (a) $T_{soil}=10°C$ and $m_{NO,ref}=0$ ppb, (b) $T_{soil}=10°C$ and $m_{NO,ref}=133$ ppb, (c) $T_{soil}=20°C$ and $m_{NO,ref}=0$ ppb NO and (d) $T_{soil}=20°C$ and $m_{NO,ref}=133$ ppb (all expressed in terms of mass of nitrogen). The transparent bands are the prediction bands of each line (95% confidence level).

3.2 NO production rates, NO consumption coefficients, and NO compensation point mixing ratios

Exemplary results of NO production rate and NO consumption coefficient as a function of gravimetric soil water content for $T_{soil}=10°C$ are shown in Fig. 6a and 6b (for soil samples taken under moss and grass cover). The NO production rate P (also expressed in ng kg^{-1} s^{-1}) is nearly as high as the net NO release rate at $m_{NO,ref}=0$ ppb. The NO production rate exponentially increased with soil moisture to a maximum value followed by a moderate decrease at higher soil moistures. This optimum shape of the NO production rate has been explained by substrate limitation under very dry conditions, and O_2-diffusion limitation under very wet conditions (Davidson et al., 1993; Meixner, 1994; Rudolph and Conrad, 1996;

Meixner and Yang, 2006; Skopp et al., 1990). The lowest optimum NO production rates were found at $T_{soil}=10°C$ for M1 and M2 with 0.7 and 0.3 ng kg^{-1} s^{-1}. G1 and G2 revealed optima of 1.2 and 1.7 ng kg^{-1} s^{-1}. S1 and S2 yield optimum NO production rates of 12.0 and 12.8 ng kg^{-1} s^{-1}, and B1 and B2 of 10.9 and 14.5 ng kg^{-1} s^{-1} (see Tab. 2). The NO production rate at 20°C showed generally higher values at optimum soil moisture. The optimum NO production rate for M1 and M2 at $T_{soil}=20°C$ were 1.1 and 0.7 ng kg^{-1} s^{-1}, for G1 and G2 3.0 and 2.0 ng kg^{-1} s^{-1}, 21.4 and 31.6 ng kg^{-1} s^{-1} for S1 and S2, and for B1 and B2 24.8 and 31.2 ng kg^{-1} s^{-1} (see Tab. 2).

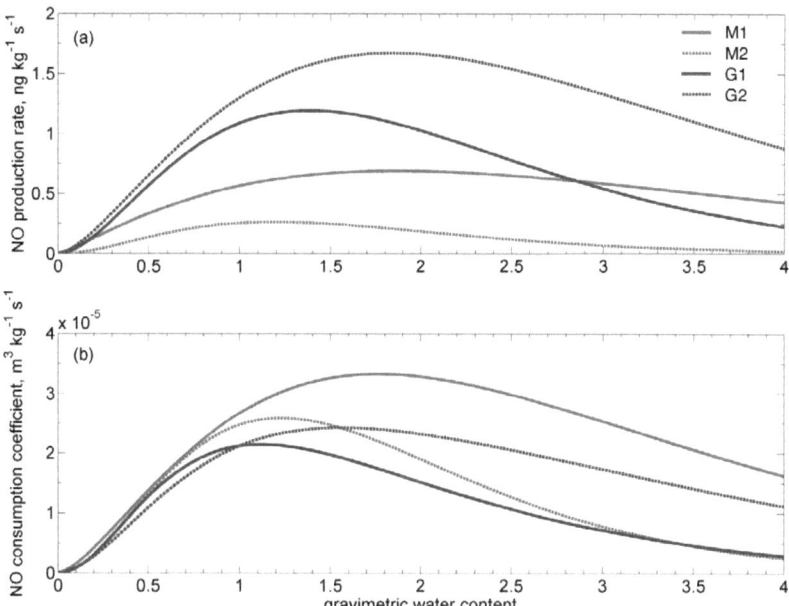

Figure 6. (a) NO production at $T_{soil}=10°C$ and (b) NO consumption coefficient at $T_{soil}=10°C$ from soil samples taken under moss and grass cover (all expressed in terms of mass of nitrogen). The red lines show the production and consumption coefficient of soil samples taken under moss covered patches and the blue lines of soil samples taken under grass covered patches.

Appendix B

The NO consumption coefficient (k, see Eq. 4) is expressed in m^3 kg^{-1} s^{-1}. For our samples, we measured maximum NO consumption coefficients for M1 and M2 of $3*10^{-5}$ m^3 kg^{-1} s^{-1}, for G1 and G2 $2*10^{-5}$ m^3 kg^{-1} s^{-1} for both, $4*10^{-5}$ m^3 kg^{-1} s^{-1} for S1 and S2 and $6*10^{-5}$ and $3*10^{-5}$ m^3 kg^{-1} s^{-1} for B1 and B2 (all values for $T_{soil}=10°C$, see Tab. 3 for NO consumption at $T_{soil}=20°C$).

Fig. 7 presents mean NO compensation point mixing ratios ($m_{NO,comp}$) for all eight soil samples at gravimetric soil moisture of 1±0.1 which is at the upper end of gravimetric soil moistures observed at the sample site (Behrendt, 2009). The $m_{NO,comp}$ varies over a wide range. Soil samples taken under moss and grass cover showed small $m_{NO,comp}$ (38 ppb and 94 ppb) compared to soil samples taken under spruce and blueberry cover which exhibited considerable higher $m_{NO,comp}$ (518 ppb and 389 ppb).

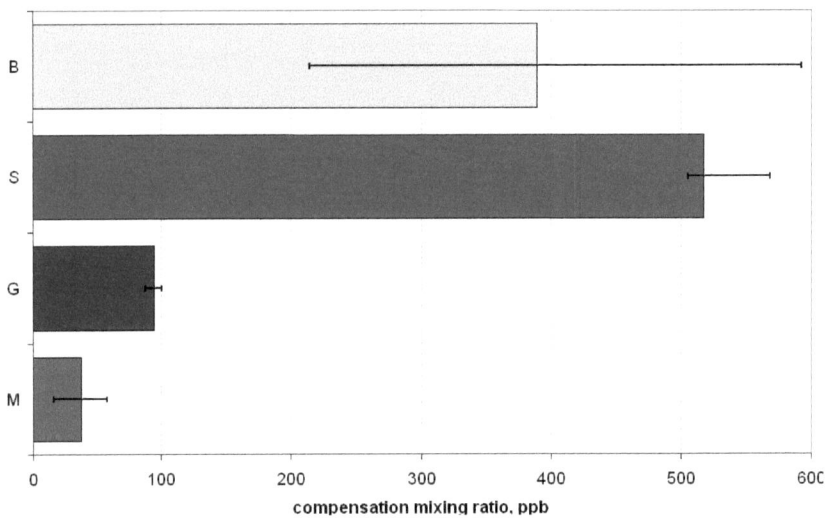

Figure 7. Median NO compensation point mixing ratios, $m_{NO,comp}$ (Eq. 7, section 2.7), for all soil probes taken under the different understory types of the Weidenbrunnen site at 1±0.1 gravimetric soil moisture and $T_{soil}=10°C$. The bars indicate the range between the 25% and 75% percentile of the data (n=10, for each understory type data set).

3.3 Net potential NO fluxes

Net potential NO fluxes derived from measured net NO release rates from soil samples taken under different understory covers are given in Fig. 8. It is remarkable, that the net potential NO fluxes from soil samples taken under spruce and blueberry cover were approximately 10-fold higher than net potential NO fluxes from soil samples taken under moss and grass cover (note different scales of y-axes in Fig. 8). The optimum NO fluxes at $T_{soil}=10°C$ ranged between 1.7 ng m^{-2} s^{-1} (M2) and 114.6 ng m^{-2} s^{-1} (B2). The position of the optimum gravimetric water content varied between the different curves. The optimum gravimetric water content for $T_{soil}=10°C$ was 0.8 for M1 and M2, 1.1 for G1 and G2, 1.3 for S1 and S2 and 1.3 for B1 and 1.5 for B2 gravimetric soil moisture (also see Tab. 2). For flushing at $T_{soil}=20°C$, optimum net potential NO fluxes were, except for S1, always higher at the higher incubation temperature. They ranged between 3.9 ng m^{-2} s^{-1} (M2) and 295 ng m^{-2} s^{-1} (B2) (see Tab. 2). Optimum gravimetric water content for $T_{soil}=20°C$ were 0.8 and 0.9 for M1 and M2, 0.5 and 0.8 for G1 and G2, 1.2 and 1.5 for S1 and S2, and 1.3 for B1 and B2.

Appendix B

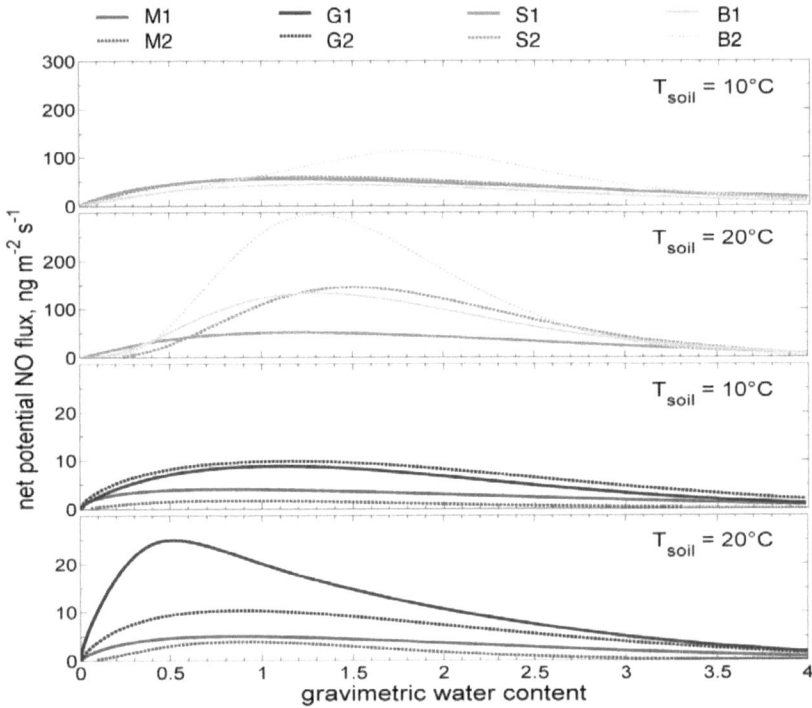

Figure 8. Net potential NO flux (all expressed in terms of mass of nitrogen) at 10°C and 20°C from soil samples taken under moss, spruce and blueberry covered patches (note different scales of the y-axes).

3.4 Temperature dependence (Q_{10} values)

Optimum net potential NO fluxes measured at two different soil temperatures (10°C and 20°C) allowed us to estimate Q_{10} values for each soil sample of the Weidenbrunnen site and data are given in Tab. 2. For S1 we derived the lowest Q_{10} value (0.92). B1 showed the highest Q_{10} value of 3.04.

Table 2. Net potential NO flux (in terms of mass of nitrogen) calculated with the diffusion coefficient according to Millington and Quirk (1960); NO production rates (P_{opt}) and the NO consumption coefficients (k_{opt}) are calculated for 10°C and 20°C and the Q_{10} values. All values are at optimum gravimetric soil moisture (θ_{opt}).

soil samples	understory vegetation	optimum gravimetric water content (10°C) [1]	optimum net potential NO flux (10°C) (ng m^{-2} s^{-1})	optimum gravimetric water content (20°C) [1]	optimum net potential NO flux (20°C) (ng m^{-2} s^{-1})	P_{opt} (10°C) (ng kg^{-1} s^{-1})	k_{opt} (10°C) (m^3 kg^{-1} s^{-1})	P_{opt} (20°C) (ng kg^{-1} s^{-1})	k_{opt} (20°C) (m^3 kg^{-1} s^{-1})	Q_{10} [1]
M1	moss	0.8	4.0	0.8	5.0	0.7	3.3×10^{-5}	1.1	5.1×10^{-5}	1.25
M2	moss	0.8	1.7	0.9	3.9	0.3	2.6×10^{-5}	0.7	3.7×10^{-5}	2.29
G1	grass	1.1	8.8	0.5	24.9	1.2	2.1×10^{-5}	3.0	3.4×10^{-5}	2.83
G2	grass	1.1	9.8	0.9	10.3	1.7	2.4×10^{-5}	2.0	3.7×10^{-5}	1.05
S1	spruce	1.3	55.4	1.2	51.1	12.0	4×10^{-5}	21.4	1.4×10^{-5}	0.92
S2	spruce	1.3	59.3	1.5	145.0	12.8	3.7×10^{-5}	31.6	4.3×10^{-5}	2.45
B1	blueberry	1.3	43.7	1.3	133.0	10.9	6.1×10^{-5}	24.8	4.2×10^{-5}	3.04
B2	blueberry	1.5	114.6	1.3	295.0	14.5	2.8×10^{-5}	31.2	2.6×10^{-5}	2.6

Appendix B

Table 3. Chemical and physical soil parameters from organic soil layers under different understories from Weidenbrunnen research site. NH_4^+ and NO_3^- are expressed in terms of mass of N.

soil samples	understory vegetation	bulk density (10^3 kg m^{-3})	particle density (10^3 kg m^{-3})	pH (measured in H$_2$O) [1]	C/N [1]	C_{org} [%]	NH_4^+ mg kg^{-1} (dry soil)	NO_3^- mg kg^{-1} (dry soil)
M1	moss	0.15	1.5	4.6	16.4	43.3	194	2
M2	moss	0.12	1.7	5	16.6	26.9	148	7
G1	grass	0.15	1.7	4.1	14.7	29.5	207	1
G2	grass	0.13	1.5	3.6	15.4	40.0	204	2
S1	spruce	0.14	1.6	3.5	16.9	43.5	56	2
S2	spruce	0.14	1.6	3.5	18.4	30.2	86	11
B1	blueberry	0.18	1.6	4.7	15.3	36.5	139	1
B2	blueberry	0.15	1.5	3.7	15.6	39.0	148	2

3.5 Chemical and physical soil parameters

The results of analysis of different soil parameters including bulk density (BD), particle density (PD), soil pH, C/N ratio, organic carbon (c_{org}), soil ammonia (NH_4^+) and soil nitrate (NO_3^-) are summarized in Tab. 3. Soil bulk density ranged between 0.12 and $0.18*10^3$ kg m^{-3}, while PD ranged between 1.5 and $1.7*10^3$ kg m^{-3}. Soil pH was lowest (3.5) in soil samples taken under spruce cover (S1, S2) and highest (5.0) for soil samples taken under moss cover (M1, M2). C/N ratios for all soil samples taken from the organic layers are relatively low, but on average (16.2) close to the range reported in literature for other Norway spruce sites in the Fichtelgebirge (see Schmitt et al., 2008; Michel et al., 2006). C/N ratios varied only in a small range, namely between 14.7 and 18.4. For C_{org} the values ranged between 26.9% (M2) and 43.5% (S1). A higher variability has been found for soil NH_4^+. Lowest soil NH_4^+ values were found for S1 (56 mg kg^{-1}) and S2 (86 mg kg^{-1}) and the highest soil NH_4^+ values were found for G1 (207 mg kg^{-1}) and G2 (204 mg kg^{-1}) (expressed in mass of N). Soil NO_3^- ranged between 1 and 11 mg kg^{-1} (expressed in mass of N).

Pearson's product-moment-analyses were performed to test (a) net potential NO fluxes (at $T_{soil}=10°C$, $T_{soil}=20°C$), (b) NO production rates (at $T_{soil}=10°C$, $T_{soil}=20°C$) and (c) NO consumption coefficients (at $T_{soil}=10°C$, $T_{soil}=20°C$) for possible relationship with the physical and chemical soil parameters (soil pH, C_{org}, C/N ratio, soil NH_4^+, soil NO_3^- and PD).

The results obtained from Pearson's product-moment-analyses are presented in Tab. 4. Significant negative correlations (probability level of 0.1) were found only between soil NH_4^+ and NO production rate at $T_{soil}=10°C$, NO production rate at $T_{soil}=20°C$, and NO consumption coefficient at $T_{soil}=20°C$. The following correlations were not significant at a probability level of 0.1. Soil pH correlated negatively with all independent variables except for the consumption coefficient at $T_{soil}=10°C$. Positive correlations with the individual variables were found for soil C_{org} and also for C/N. Soil NO_3^- vs. independent variables showed positive correlations except for the NO consumption coefficient. Particle density correlated negatively with independent variables, except for the NO consumption coefficient.

Appendix B

Table 4. Results of Pearson product moment correlation analysis of net NO release rates, net potential NO flux, NO production rate (P) and NO consumption coefficient (k) versus physical and chemical soil parameters

	NO flux 10°C	P 10°C	k 10°C	NO flux 20°C	P 20°C	k 20°C
pH	-0.537	-0.523	0.206	-0.356	-0.468	-0.311
C_{org}	0.226	0.213	0.227	0.066	0.095	0.495
C/N	0.157	0.332	0.137	0.043	0.364	0.325
NH_4^+	-0.519	-0.739[a]	-0.467	-0.315	-0.698[a]	-0.662[a]
NO_3^-	0.061	0.172	-0.033	0.091	0.274	-0.083
PD	-0.367	-0.181	0.023	-0.387	-0.149	0.171

[a]Significant at the 0.10 probability level

4 Discussion

4.1 Comparison with other studies

During the last two decades, there has been a series of studies on biogenic NO emissions from soil in forest ecosystems (Papke and Papen, 1998; Pilegaard et al., 2006; Kesik et al., 2005; Johansson, 1984; Pilegaard et al., 1999; Butterbach-Bahl et al., 2002; Lehmann, 2002; Butterbach-Bahl et al., 2001). However, there are only a few studies examining spatial differences of NO fluxes within a forest (Gasche and Papen, 1999; Lehmann, 2002; Pilegaard et al., 1999; Nishina et al., 2009). Furthermore, the influence of soil organic matter on soil biogenic NO emissions has not been studied in detail and is consequently not well known. In most studies the effect of the dominant overstory or of the whole soil core (mineral and

Appendix B

organic layer) was addressed by measurements using the dynamic chamber technique (Gasche and Papen, 1999; Butterbach-Bahl et al., 1997; Johansson, 1984).

During the last two decades, a series of field and laboratory studies clearly demonstrated, that NO fluxes, measured in the field by dynamic chamber techniques, were in good agreement with those NO fluxes, which have been derived from laboratory incubations on soils sampled from the top soil layer of dynamic chambers' enclosures (Meixner et al., 1997; van Dijk et al., 2002; Remde et al., 1993; Ludwig et al., 2001; Meixner and Yang, 2006; Otter et al., 1999). However, for more detailed investigations, laboratory studies are necessary, but only a few groups seem to have the facilities available to carry out laboratory measurements of soil NO exchange (e.g. Bollmann et al., 1999; Ormeci et al., 1999; Schindlbacher et al., 2004; Feig et al., 2008; van Dijk and Meixner, 2001). Since laboratory studies are outnumbered, most of the following discussion is based on results from field measurements in spruce forests.

Pilegaard et al. (1999), applying a dynamic field chamber technique in a spruce forest site at Ulborg (Denmark), found low NO fluxes from moss covered soil. However, NO fluxes increased with closeness to standing tree trunks. For their forest soils which had a thick organic layer (4 cm), NO fluxes ranged between <0.3 and 66 ng m^{-2} s^{-1}. Similar results were presented by Gasche and Papen (2002) for the Höglwald forest (Germany). Their measurements, also employing a dynamic chamber technique, addressed the spatial distribution of NO fluxes along a tree-to-tree gradient. For 1997, annual mean NO fluxes of 29.2±0.9 ng m^{-2} s^{-1} were found for those chambers which were located closest to the stems, 18.4±0.5 ng m^{-2} s^{-1} for chambers approx. 4 m, and 12.3±0.4 ng m^{-2} s^{-1} for the chamber approx. 6 m apart from the stems. With closeness to trunks (living trees) the NO emissions increased significantly (between 1.6- and 2.6-fold). While for the Höglwald beech forest site, Gasche and Papen (2002) could explain an identical spatial effect with marked differences in soil physical and chemical soil parameters, there was no detailed explanation for the Höglwald spruce forest site. One reason could be that the nutrient supply from stem flow is negligible at this spruce forest site (Gasche and Papen, 2002). Butterbach-Bahl et al. (1997) reported mean monthly NO fluxes between 5.6 and 36.1 ng m^{-2} s^{-1} for the same Höglwald spruce forest site (July 1994 to June 1995). The site exhibits acidic soil pH values (2.7 to 3.6) in the organic layer. Again for the Höglwald spruce site, Gasche and Papen (1999) showed, that most of the

NO emissions came from the organic layer and only a small contribution from the mineral soil. For the entire Höglwald site, they reported annual NO emission rates of 25.5±0.5 ng m^{-2} s^{-1} during 1994-1996. Very low NO fluxes (0.3±0.1 ng m^{-2} s^{-1}) were reported by Horváth et al. (2006) for a spruce forest site in NE Hungary (October 2002 to September 2003). Similar low NO fluxes were reported by Kitzler et al. (2006) for the spruce-fir-beech forest site of Achental (Austria) during the period of May 2002 to July 2004. Using a dynamic chamber technique, they found mean NO fluxes of only 0.2±0.02 ng m^{-2} s^{-1} for the first year and mean NO fluxes of 0.14±0.01 ng m^{-2} s^{-1} for the second year. However, the pH values at this site are very high (6.42).

Laboratory studies on undisturbed soil samples from the Weidenbrunnen site (approx. 300 m west of our site) resulted in NO fluxes between 2.6 and 12.9 ng m^{-2} s^{-1} (Muhr et al., 2008). This site is also a spruce site mainly covered with grass. Another laboratory study on mineral soil samples (taken just from the A horizon) were carried out at the Nagoya University Forest (Japan) site covered with Japanese cedar. The NO emissions ranged from 0.3 ng m^{-2} s^{-1} at high soil water contents (<92% WFPS) to 72.2 ng m^{-2} s^{-1} at low soil water contents (>29% WFPS)(Nishina et al., 2009).

Our optimum net potential NO fluxes for soil samples taken under grass cover (8.8-9.8 ng m^{-2} s^{-1}, T_{soil}=10°C, see Tab. 2) agree well with the (laboratory) results of Muhr et al. (2008). Also the results of Nishina et al. (2009) are in the range of our optimum net potential NO fluxes. However, their soil samples were taken from the mineral soil. Our results for soil samples taken under grass cover also overlap with the data given by Butterbach-Bahl et al. (1997) and Gasche and Papen (2002). Annual NO emission rates measured by Gasche and Papen (1999) range between our optimum net potential NO fluxes for soil samples taken under grass and spruce cover (and also for the B1 sample). Contrastingly, the NO fluxes found by Kitzler et al. (2006) and Horváth et al. (2006) are much lower than any of our optimum net potential NO fluxes. However, the Achental site is a mixed forest, and the soil exhibits a relatively high pH value (6.42). Relatively high values of the soil water content (average: 53% WFPS) characterized the soils of the Hungarian site (see Horváth et al., 2006). In contrast, the optimum soil water contents found in our study ranged between 18 and 27% WFPS (see equivalent gravimetric water contents in Tab. 2). Optimum net potential NO fluxes of our moss covered soils (if watered to 53% WFPS) would fall in the range of field fluxes observed by Horváth et al. (2006). In any case, our optimum net potential NO fluxes from soils under spruce and blueberries show higher values than any fluxes of the other studies mentioned above.

Net potential NO fluxes derived from laboratory experiments using the algorithm of Galbally and Johansson (1989) are particularly sensitive to changes in NO production rates and NO consumption coefficients and less sensitive to changes in diffusivity and soil bulk density (Rudolph and Conrad, 1996). In this respect, when comparing NO soil flux estimates (derived from laboratory incubation measurements), with data from literature, one should keep in mind, that the most up-to-date diffusion coefficient equations are basically applicable only to mineral soils. As gas diffusion in the organic layer can be substantially different, and uncertainties in determining diffusion coefficients in organic layers are still a matter of discussion (Moldrup, personal communication), we employed different mathematical formulations (see Tab. 1), and found that the choice of the diffusion coefficient equation had an effect on the calculated NO flux (see Fig. 3). Depending on the diffusion coefficient, the NO fluxes had different magnitudes (factor of maximum 1.26 over the entire soil moisture range) and exhibited a shift in the position of the optimum flux (see Fig. 3). However, even using the correct effective diffusion coefficient, attention should be paid to its determination as the equation includes both the bulk and particle density. Both densities vary significantly between organic and mineral soil layers (e.g. Weidenbrunnen site: organic soil layers: BD: 0.14±0.02, PD: 1.6±0.07, n=8, mineral soil layers: BD: 0.88±0.18, PD: 2.47±0.06, n=8). If the effective diffusion coefficient has to be calculated, it is necessary to measure these quantities directly. Nevertheless, to reveal the uncertainties in diffusion through organic soil layers, further research, especially through field measurements of the diffusion coefficient, are most desirable.

Comparisons of NO production rates are not affected by the choice of diffusion coefficients. Therefore, only a few NO production rates are reported in the literature. Venterea and Rolston (2000) found mean NO production rates in a range of 9.4 to 18.7 ng kg^{-1} s^{-1} for agricultural soils from the Sacramento Valley of California. These values are comparable with our results of NO production rates (0.3-14.5 ng kg^{-1} s^{-1}). Remde et al. (1989) reported NO production

Appendix B

rates twice as high as ours for a sandy clay loam under aerobic conditions (27.4 ± 1.8 ng kg^{-1} s^{-1}), yet much higher under anaerobic conditions (738 ± 21.6 ng kg^{-1} s^{-1}).

NO production rates reported in the literature are as rare as NO consumption coefficients. Values of the NO consumption coefficient (k) found in this study were in the range of $2*10^{-5}$ to $6*10^{-5}$ m^3 s^{-1} kg^{-1}. Soils from the Bolivian Amazon region showed k values under oxic conditions of $8*10^{-5}$ m^3 s^{-1} kg^{-1} (Koschorreck and Conrad, 1997). Feig et al. (2008) reported NO consumption coefficients between $5*10^{-5}$ and $26*10^{-5}$ m^3 s^{-1} kg^{-1}. However, these values were determined for desert soil with nearly no organic material inside at 25°C in the laboratory, 15°C more than for our studies. As biological processes usually increase by a factor of two with an increase in temperature of 10°C (Kirschbaum, 1995; Davidson et al., 2006; Zheng et al., 2003), the higher values reported by Feig et al. (2008) are to be expected.

Only a few studies reported compensation point mixing ratios. Slemr and Seiler (1991) determined NO compensation point mixing ratio for agricultural soils between 0.3 and 5.5 ppb. Gasche and Papen (1999) found NO compensation point mixing ratios of 69.9 ± 9.6 ppb for a spruce forest soil in the Höglwald, Germany. Only for G1 and G2 we found NO compensation points in the low range of these studies. Soil samples taken under spruce and blueberry cover showed a much higher $m_{NO,comp}$. However, there are also studies which found higher NO compensation mixing ratios, e.g. ranging between 9 and 875 ppb for agriculture, meadow and forest soils (Gödde and Conrad, 2000). In view of the ambient NO mixing ratios observed at the Weidenbrunnen site, NO compensation point mixing ratios found in our study demonstrate, that the soils there mainly act as a biogenic source for NO. Only when the ambient NO mixing ratio matches or falls below the NO compensation point mixing ratio will the soils become a sink for biogenic NO. Moravek (2008) observed ambient NO mixing ratios between 1 and 2 ppb at 5 cm above the forest floor (moss covered) and Plake (2009) found NO mixing ratios up to 4.2 ppb at 0.5 cm above the forest floor (moss covered), both at the Weidenbrunnen site. These mixing ratios are too low to change the NO flux from upward to downward directions.

Many studies have presented an exponential increase of soil NO emissions with increasing temperature. Generally, Q_{10} values are in the range of 2-3, a range valid for most biochemical processes (Koponen et al., 2006; Kirkman et al., 2002; van Dijk et al., 2002; Feig et al., 2008;

Meixner and Yang, 2006; Smith et al., 2003). We obtained Q_{10} values for net potential NO fluxes between 0.92 and 3.04 (see Tab. 3). However, a Q_{10} value less than 1 (e.g. S1: 0.92) indicates a decrease of soil NO emission with increasing temperature.

4.2 Influence of soil chemical parameters on net potential NO flux

The processes which result in NO exchange are mainly influenced by soil temperature and soil moisture (Davidson and Kingerlee, 1997; Johansson and Granat, 1984; Skiba et al., 1997; Ludwig et al., 2001; Feig et al., 2008; Meixner, 1994; Meixner and Yang, 2006). Nevertheless, soil chemical and physical parameters may also affect the NO exchange (Nägele and Conrad, 1990a; Smith et al., 2003; Ludwig et al., 2001; Pilegaard et al., 2006; Kitzler et al., 2006; Laville et al., 2009; Gödde and Conrad, 2000).

For our soil samples from the Weidenbrunnen site we found no significant (probability level of 0.05) relationships between optimum net potential NO fluxes, NO production rates, or NO consumption coefficients with any physical or chemical soil parameters. However, on the 0.1 significance level we found negative correlations between soil NH_4^+ and (a) NO production rate (at $T_{soil}=10°C$), (b) NO production rate (at $T_{soil}=20°C$), and (c) NO consumption coefficient (at $T_{soil}=20°C$) (see Tab. 4). Also the NO consumption coefficient at $T_{soil}=10°C$ and the net potential NO fluxes showed a negative, but not significant correlation with soil NH_4^+. These negative correlations with soil NH_4^+ point to nitrification as the main converting process, because soil NH_4^+ must be available before nitrification may start. Denitrification is the conversion of NO_3^- to N_2O or N_2, and NO_3^- is necessary for the activation of denitrification. However, denitrification seems to play a smaller role for soils from the Weidenbrunnen site because we found no significant correlation between soil NO_3^- and other variables (see Tab. 4). Furthermore, nitrification may be lower from soil samples taken under moss and grass than from soil samples taken under spruce and blueberry cover. Therefore, the amount of soil NH_4^+ is higher at soil samples taken under moss and grass than at soil samples taken under spruce and blueberry cover. Gödde and Conrad (2000) also found, that nitrification is the dominant process of NO production in the soil. In contrast to our study, Baumgärtner and Conrad (1992) found no significant correlation between the NO production rate and soil NH_4^+, but did find a significant correlation between the NO consumption coefficient and soil NH_4^+. However, they investigated mineral soil only. NO production and NO consumption processes are differently regulated (Dunfield and Knowles, 1998), so that both processes can respond independantly to changes in external factors. Gasche and Papen (1999) found a correlation between NO fluxes and soil NH_4^+ for the Höglwald spruce forest site as well as a correlation between NO fluxes and soil NO_3^-. Typically, 1-4% (sometimes more) of soil NH_4^+ is released from soil as NO (Dunfield and Knowles, 1998).

The other parameters showed no significant correlations (<0.1). As the research site, a typical even-aged monoculture, is relatively small (1.4 ha), soil parameters vary only over a small range (see also Behrendt, 2009). This makes it difficult or impossible to establish significant correlations between the other soil chemical or physical parameters and net potential NO fluxes.

Nevertheless, net potential NO fluxes showed a weak relationship with soil pH values. During laboratory incubation measurements, there might have been microsites in the soil samples with a soil pH different from the measured mean pH, indicating that nitrification occurred in microsites having pH higher than the surrounding soil (Paavolainen and Smolander, 1998). That could also be a reason for the relatively high NO emission despite of the low pH values. A pH value between 7 and 8 is ideal for nitrification. However, Paavolainen and Smolander (1998) reported coniferous soils that exhibited acid-tolerant nitrification. In this respect, a series of studies reported relationships between NO exchange processes and soil pH (Gödde and Conrad, 2000; Venterea et al., 2004; Nägele and Conrad, 1990b). There is also an enhanced chemical NO production from nitrite at low soil pH (Cleemput and Baert, 1984), which can happen even if nitrite does not accumulate to detectable amounts. In contrast, other studies found no strong relationships between NO exchange and soil pH (Dunfield and Knowles, 1998).

4.3 Influence of the understory type on net potential NO flux

A number of studies have detected effects of vegetation on NO emissions (Meixner et al., 1997; Feig et al., 2008; Davidson, 1991; Martin and Asner, 2005; Pilegaard et al., 1999). Our study suggests a strong relationship between understory type and the amount of net potential NO flux. As this relationship can hardly be explained by the measured physical and chemical soil parameters alone, it may originate from the complex biological interactions between plants and their soil environment. Because plant species differ in quantity and quality of resources that they return to soil, individual plant species may have important effects on components of the soil biota and the processes that they regulate (Wardle et al., 2004). Carbon derived from plant litter mainly influences the decomposer communities. In addition, providing carbon to the rhizosphere creates a hot spot for microbial activity in the soil. For example the size and the density of the nitrifier and denitrifier communities are strongly influenced by plant roots (Philippot et al., 2009).

In our experiments, soils were sieved through a 16 mm sieve and kept at 4°C for up to 2 months. Sieving may have removed the majority of roots but it cannot be excluded that fine roots passed the meshes resulting in a soil sample containing litter, roots, rhizosphere and root free soil. Stöhr and Ullrich (2002), and Stöhr and Stremlau (2006) demonstrated that roots can generate NO. The contribution of living roots to the observed net potential NO fluxes in our experiments should be rather low because most fine roots were removed by sieving. However, biochemical reactions of intact fine roots of spruce when stored in soil at 4°C are unchanged for up to 4 weeks and then slowly decline (Pritsch, unpublished results). Thus it cannot be excluded that a minor part of the observed NO emissions came directly from those fine roots that were not removed by sieving.

A more likely explanation for the different net potential NO fluxes is that litter type and the influence of root exudates influenced functions of the soil microbial communities under the respective understory plants. Rhizosphere effects i.e. the influence of roots on NO emission rates was found by Slemr and Seiler (1991). Vos et al. (1994) measured 2 to 12-fold higher NO emissions from plots covered with green manure than from fallow plots, probably caused by increased microbial activity in the rhizosphere of the green manure plots compared to the bare soil. Unfortunately, no field studies exist examining the influence of plant roots on NO emissions. A few studies have shown a strong influence of roots on nitrous oxide emissions (Mosier et al., 1990) and it is generally accepted that denitrification is highest in the rhizosphere and decreases with distance from plant roots (Smith and Tiedje, 1979).

According to our study, net potential NO fluxes as well as NO production rates, NO consumption coefficients, and net NO release rates displayed the highest values for soil samples taken under spruce and blueberry covered soils and the lowest values for soil samples taken under moss and grass covered soils. Our results on small net potential NO fluxes from soils taken under moss cover are in accordance with findings of Pilegaard et al. (1999). They suspected that mosses retain nutrients from throughfall but also hypothesized that moss cover simply reflects other factors such as canopy density and water availability. Similarly small net potential NO fluxes were found for soil samples collected under grass cover in our study. *Deschampsia flexuosa* has a high potential to take up nitrogen in various forms and in competition to microbes (Harrison et al., 2008). This may explain a possibly reduced potential of its microbial communities in nitrogen cycling. The role of its arbuscular mycorrhizal (AM) associates has not been studied at the field site but colonisation by AM seems to be low on acidic soils (Göransson et al., 2008). Inferior competition of microbial communities under moss and grass cover therefore could explain low NO emissions.

Soils taken under blueberry and spruce cover, in contrast, produced high net potential NO fluxes. Both plant species are associated with asco- and basidiomycetes forming ericoid mycorrhizae (blueberry), respectively ectomycorrhizae (spruce). NO accumulation can occur in mycorrhizal symbioses (Stöhr and Stremlau, 2006). Wallenda et al. (2000) also demonstrated that intact mycorrhizal roots of Norway spruce took up substantial amounts of NH_4^+. This NH_4^+ may act as precursor of nitrification. During nitrification NO can be released as an intermediate. However, due to the fact that only very few roots may have been present and in an active state NO released from mycorrhizae may be of minor relevance. The 10 fold higher NO fluxes from the soils beneath spruce and blueberry are difficult to explain from our data. One factor may be that both plants produce litter types rich in lignin and phenolics (Adamczyk et al., 2008). Tannins formed in degradation of these litter types can form complexes with proteins. Protein phenol complexes can be degraded by ericoid mycorrhizal fungi and saprotrophic fungi but not by ectomycorrhizal fungi (Wu et al., 2003). It has been suggested that relatively more dissolved organic nitrogen (DON) compared to inorganic nitrogen is released upon degradation of these phenol rich litters (cf. from (Hofland-Zijlstra and Berendse, 2010). Since DON as a possible substrate for nitrification and N-mineralisation has not been measured in our study it can only be speculated if nitrogen sources other than NH_4^+ could explain the high NO net release or which part of the soil microflora may have contributed to the results. It could be speculated that fungi as decomposers may have played a

Appendix B

role in this process. In a beech forest, measurements of nitrous oxide (N_2O) emission from forest floor samples indicated that net N_2O production was the result of predominantly fungal N_2O production and predominantly bacterial N_2O consumption (Blagodatskaya et al., 2010).

Altogether our results indicate a challenging field for unravelling the underlying processes of different understory plants on NO net release from forest soils.

5 Conclusion

In this study, we investigated the net potential NO fluxes from soil samples of the organic layers of a spruce forest soil covered with four different understory types (moss, grass, spruce and blueberry).

Observed net NO release rates of soil samples taken under moss and grass cover indicated a high potential for NO consumption, resulting in very low net potential NO fluxes from soil samples taken under these understory types. In strong contrast, soil samples taken under spruce and blueberry cover showed 10 fold higher net potential NO fluxes, than those taken under moss and grass cover.

Therefore, it is an important lesson of this study, that more attention must be paid to small scale heterogeneity of understory vegetation, when quantification of the biogenic NO emission from a (spruce) forest floor is attempted.

Analysis of the compensation point mixing ratios indicated that measured ambient mixing ratios of NO at 0.5 cm above the forest floor of the field site were – even for the soil samples taken under moss and grass cover – too low to change the soil NO flux from upward to downward directions.

Further research investigating effective soil diffusion coefficients is very desirable. The net potential NO flux calculated with the diffusion coefficient according to Millington (1959) is 1.26 fold higher than the net potential NO flux calculated with diffusion coefficients according to Millington and Quirk (1960). Also the position of the optimum NO flux shifts depending on the choice of the diffusion coefficient.

Appendix B

While the understory type seems to be an important variable controlling NO exchange processes, corresponding soil nutrients played generally a less important role. The only exception was for NH_4^+, the precursor of NO_3^- in the nitrification process. This implies that nitrification was the limiting factor of NO production for the investigated soils, whereas denitrification played an obviously smaller role. It is remarkable that high NO emissions were observed for soils under woody understory types; this may be related to soil chemical processes in the vicinity of mycorrhized roots, but further studies are certainly necessary for confirmation. As the establishment of different understory types is related to the availability of light at the forest floor as a result of forest thinning, management practises are likely to have important consequences on the net soil NO emission from a forested site.

Coniferous forest soils in temperate humid climates are characterized by thick organic layers of moder or raw humus forms. Organic layers of our soils had a much higher potential (over 2.5 fold) for NO emission than the corresponding mineral soil layers. Hence quantification of net potential NO fluxes of the O horizons of temperate forest soils is an important step for (a) comparison of laboratory and field measurements, (b) up-scaling from laboratory to field scale fluxes (by areal information on understory distribution), and (c) extrapolation from field site results to larger scales (e.g. regional).

Acknowledgements

The authors gratefully acknowledge financial support by the German Science Foundation (DFG project: EGER - "ExchanGE processes in mountainous Regions" (ME 2100-4)) and by the Max Planck Society. Thomas Behrendt is thanked for collecting the soil samples. Particle density analyses were performed at the Landesamt für Geologie und Bergbau in Mainz, Germany. Soil NO_3^-, soil NH_4^+, and soil pH were analysed through Bayreuth Center of Ecology and Environmental Research (BayCEER), University Bayreuth, Germany. C_{org} and C/N ratio were analysed through the laboratory of the Institute of Geography, University of Mainz, Germany.

Appendix B

References

Adamczyk, B., Kitunen, V., and Smolander, A.: Protein precipitation by tannins in soil organic horizon and vegetation in relation to tree species, Biology and Fertility of Soils, 45, 55-64, 10.1007/s00374-008-0308-0, 2008.

Baumgärtner, M., and Conrad, R.: Effects of soil variables and season on the production and consumption of nitric oxide in oxic soils, Biology and Fertility of Soils, 14, 166-174, 1992.

Behrendt, T.: A small-scale geostatistical analysis of the variability of soil properties driving the biogenic emission of nitric oxide from soil, MSc thesis, Geography, Johannes Gutenberg University Mainz, Mainz, Germany, 2009.

Blagodatskaya, E., Dannenmann, M., Gasche, R., and Butterbach-Bahl, K.: Microclimate and forest management alter fungal-to-bacterial ratio and N2O-emission during rewetting in the forest floor and mineral soil of mountainous beech forests, Biogeochemistry, 97, 55-70, 2010.

Bollmann, A., and Conrad, R.: Influence of O_2 availability on NO and N_2O release by nitrification and denitrification in soils, Global Change Biology, 4, 387-396, 1998.

Bollmann, A., Koschorreck, M., Meuser, K., and Conrad, R.: Comparison of two different methods to measure nitric oxide turnover in soils, Biology and Fertility of Soils, 29, 104-110, 1999.

Butterbach-Bahl, K., Gasche, R., Breuer, L., and Papen, H.: Fluxes of NO and N_2O from temperate forest soils: impact of forest type, N deposition and of liming on the NO and N_2O emissions, Nutrient Cycling in Agroecosystems, 48, 79-90, 1997.

Butterbach-Bahl, K., Stange, F., and Papen, H.: Regional inventory of nitric oxide and nitrous oxide emissions for forest soils of southeast Germany using the biogeochmical model PnET-N-DNDC, Journal of Geophysical Research, 106, 34,155-134,166, 2001.

Butterbach-Bahl, K., Rothe, A., and Papen, H.: Effect of tree distance on N_2O and CH_4-fluxes from soils in temperate forest ecosystems, Plant and Soil, 240, 91-103, 2002.

Cleemput, O., and Baert, L.: Nitrite: a key compound in N loss processes under acid conditions?, Plant and Soil, 76, 233-241, 1984.

Conrad, R.: Compensation Concentration as Critical Variable for Regulating the Flux of Trace Gases between Soil and Atmosphere, Biogeochemistry, 27, 155-170, 1994.

Crutzen, P. J.: Role of NO and NO_2 in the chemistry of the troposphere and stratosphere, Annual Review of Earth and Planetary Sciences, 7, 443-472, 1979.

Davidson, E. A.: Fluxes of nitrous oxide and nitric oxide from terrestrial ecosystems, in: Microbial production and consumption of greenhouse gases: methane, nitrogen oxides, and halomethanes, edited by: Rogers, J. E., and Whitman, W. B., American Society for Microbiology, Washington D.C., 219-235, 1991.

Davidson, E. A., Matson, P. A., Vitousek, P. M., Riley, R., Dunkin, K., Garcia-Mendez, G., and Maass, J. M.: Processes regulating soil emissions of NO and N_2O in a seasonally dry tropical forest, Ecology, 74, 130-139, 1993.

Davidson, E. A., and Kingerlee, W.: A global inventory of nitric oxide emissions from soils Nutrient Cycling in Agroecosystems, 48, 37-50, 1997.

Appendix B

Davidson, E. A., Janssens, I. A., and Luo, Y.: On the variability of respiration in terrestrial ecosystemy: moving beyond Q_{10}, Global Change Biology, 12, 154-164, 2006.

Denman, K. L., Brasseur, G. P., Chidthaisong, A., Ciais, P., Cox, P. M., Dickinson, R. E., Hauglustaine, D., Heinze, C., Holland, E. A., Jacob, D. J., Lohmann, U., Ramachandran, S., da Silva Dias, P. L., Wofsy, S. C., and Zhang, X.: Couplings between changes in the climate system and boigeochemistry, in: Climate Change 2007: The physical science basis. contribution of working group 1 to the fourth assesment report of the Intergovernmental Panel on Climate Change, edited by: Solomon, S., Qin, D., Manning, M., Chen, Z., Marquis, M., Averyt, K. B., Tignor, M., and Miller, H. L., University Press Cambridge, 2007.

Dunfield, P. F., and Knowles, R.: Organic matter, heterotrophic activity, and NO consumption in soils, Global Change Biology, 4, 199-207, 1998.

Falge, E., Tenhunen, J. D., and Aubinet, M.: A model-based study of carbon fluxes at ten European forest sites, in: Fluxes of carbon, water and energy of European forests edited by: Valentini, R., Springer, Berlin, 151-177, 2003.

Feig, G. T., Mamtimin, B., and Meixner, F. X.: Soil biogenic emissions of nitric oxide from a semi-arid savanna in South Africa, Biogeosciences, 5, 1723-1738, 2008.

Firestone, M. K., and Davidson, E. A.: Microbiological basis of NO and N_2O production and consumption in soil in: Exchange of trace gases between terrestrial ecosystems and the atmosphere, edited by: Andreae, M. O., and Schimel, D. S., Wiley, Chichester, 7-21, 1989.

Foken, T.: Lufthygienisch-biologische Kennzeichnung des oberen Egertales (Fichtelbegirge bis Karlovy Vary), Bayreuther Forum Ökologie, 100, 1-70, 2003.

Fowler, D., Pilegaard, K., Sutton, M. A., Ambus, P., Raivonen, M., Duyzer, J., Simpson, D., Fagerli, H., Fuzzi, S., Schjoerring, J. K., Granier, C., Neftel, A., Isaksen, I. S. A., Laj, P., Maione, M., Monks, P. S., Burkhardt, J., Daemmgen, U., Neirynck, J., Personne, E., Wichink-Kruit, R., Butterbach-Bahl, K., Flechard, C., Tuovinen, J. P., Coyle, M., Gerosa, G., Loubet, B., Altimir, N., Gruenhage, L., Ammann, C., Cieslik, S., Paoletti, E., Mikkelsen, T. N., Ro-Poulsen, H., Cellier, P., Cape, J. N., Horváth, L., Loreto, F., Niinemets, Ü., Palmer, P. I., Rinne, J., Misztal, P., Nemitz, E., Nilsson, D., Pryor, S., Gallagher, M. W., Vesala, T., Skiba, U., Brüggemann, N., Zechmeister-Boltenstern, S., Williams, J., O'Dowd, C., Facchini, M. C., de Leeuw, G., Flossman, A., Chaumerliac, N., and Erisman, J. W.: Atmospheric composition change: Ecosystems-Atmosphere interactions, Atmospheric Environment, 43, 5193-5267, 2009.

Galbally, I. E.: Factors controlling NO_x emissions from soils, in: Exchange of trace gases between terrestrial ecosystems and the atmosphere, edited by: Andreae, M. O., and Schimel, D. S., Wiley & Sons, Chichester, 23-37, 1989.

Galbally, I. E., and Johansson, C.: A model relating laboratory measurements of rates of nitric-oxide production and field-measurements of nitric-oxide emission from soils, Journal of Geophysical Research-Atmospheres, 94, 6473-6480, 1989.

Gasche, R., and Papen, H.: A 3-year continuous record of nitrogen trace gas fluxes from untreated and limed soil of a N-saturated spruce and beech forest ecosystem in Germany 2. NO and NO_2 fluxes, Journal of Geophysical Research, 104, 505-520, 1999.

Gasche, R., and Papen, H.: Spatial variability of NO and NO_2 flux rates from soil of spruce and beech forest ecosystems, Plant and Soil, 240, 67-76, 2002.

Gelfand, I., Feig, G., Meixner, F. X., and Yakir, D.: Afforestation of semi-arid shrubland reduces biogenic NO emission from soil, Soil Biology & Biochemistry, 41, 1561-1570, 2009.

Gerstenberger, P., Foken, T., and Kalbitz, K.: The Lehstenbach and Steinkreuz catchment in NE Bavaria, Germany, in: Biogeochemistry of forested catchments in a changing environment: a german case study, edited by: Matzner, E., Springer, Berlin, 15-44, 2004.

Gödde, M., and Conrad, R.: Influence of soil properties on the turnover of nitric oxide and nitrous oxide by nitrification and denitrification at constant temperature and moisture, Biology and Fertility of Soils, 32, 120-128, 2000.

Göransson, P., Olsson, P. A., Postma, J., and Falkengren-Grerup, U.: Colonisation by arbuscular mycorrhizal and fine endophytic fungi in four woodland grasses - variation in relation to pH and aluminium, Soil Biology & Biochemistry, 40, 2260-2265, 10.1016/j.soilbio.2008.05.002, 2008.

Guthrie, T. F., and Duxbury, J. M.: Nitrogen Mineralization and denitrification in organic soil, Soil Sci Soc Am J, 42, 908-912, 1978.

Harrison, K. A., Bol, R., and Bardgett, R. D.: Do plant species with different growth strategies vary in their ability to compete with soil microbes for chemical forms of nitrogen?, Soil Biology & Biochemistry, 40, 228-237, 10.1016/j.soilbio.2007.08.004, 2008.

Hofland-Zijlstra, J. D., and Berendse, F.: Effects of litters with different concentrations of phenolics on the competition between Calluna vulgaris and Deschampsia flexuosa, Plant and Soil, 327, 131-141, 10.1007/s11104-009-0037-7, 2010.

Horváth, L., Führer, E., and Lajtha, K.: Nitric oxide and nitrous oxide emission from Hungarian forest soils; linked with atmospheric N-deposition, Atmospheric Environment, 40, 7786-7795, 2006.

Jambert, C., Delmas, R. A., Labroue, L., and Chassin, P.: Nitrogen compound emissions from fertilized soils in a maize field pine tree forest agrosystem in the southwest of France, Journal of Geophysical Research-Atmospheres, 99, 16,523-516,530, 1994.

Johansson, C.: Field measurements of emission of nitric oxide from fertilized and unfertilized forest soils in Sweden, Journal of Atmospheric Chemistry, 1, 429-442, 1984.

Johansson, C., and Granat, L.: Emission of nitric oxide from arable land, Tellus, 36B, 25-37, 1984.

Kapiluto, Y., Yakir, D., Tans, P., and Berkowitz, B.: Experimental and numerical studies of the O-18 exchange between CO_2 and water in the atmosphere-soil invasion flux, Geochimica Et Cosmochimica Acta, 71, 2657-2671, 2007.

Kesik, M., Ambus, P., Baritz, R., Bruggemann, N. B., Butterbach-Bahl, K., Damm, M., Duyzer, J., Horvath, L., Kiese, R., Kitzler, B., Leip, A., Li, C., Pihlatie, M., Pilegaard, K., Seufert, G., Simpson, D., Skiba, U., Smiatek, G., Vesala, T., and Zechmeister-Boltenstern, S.: Inventories of N_2O and NO emissions from European forest soils, Biogeosciences, 2, 353-375, 2005.

Kirkman, G. A., Gut, A., Ammann, C., Gatti, L. V., Cordova, A. M., Moura, M. A. L., Andreae, M. O., and Meixner, F. X.: Surface exchange of nitric oxide, nitrogen dioxide, and ozone at a cattle pasture in Rondonia, Brazil, Journal of Geophysical Research-Atmospheres, 107, 10.1029/2001JD000523, 2002.

Kirschbaum, M. U. F.: The temperature dependence of soil organic matter decomposition, and the effect of global warming on soil organic C storage, Soil Biology and Biochemistry, 27, 753-760, 1995.

Kitzler, B., Zechmeister-Boltenstern, S., Holtermann, C., Skiba, U., and Butterbach-Bah, K.: Controls over N_2O, NO_x and CO_2 fluxes in a calcareous mountain forest soil, Biogeosciences, 3, 383-395, 2006.

Koponen, H. T., Escudé Duran, C., Maljanen, M., Hytönen, J., and Martikainen, P. J.: Temperature responses of NO and N_2O emissions from boreal organic soil, Soil Biology and Biochemistry, 38, 1779-1787, 2006.

Koschorreck, M., and Conrad, R.: Kinetics of nitric oxide consumption in tropical soils under oxic and anoxic conditions, Biology and Fertility of Soils, 25, 82-88, 1997.

Laville, P., Flura, D., Gabrielle, B., Loubet, B., Fanucci, O., Rolland, M. N., and Cellier, P.: Characterisation of soil emissions of nitric oxide at field and laboratory scale using high resolution method, Atmospheric Environment, 43, 2648-2658, 2009.

Lehmann, L.: Untersuchungen zu biogenen NO-Emissionen aus einem mitteleuropäischen Laubwaldboden, Msc thesis, Department of Geography and Geoecology, University of Kalsruhe, Karlsruhe, Germany, 92 pp., 2002.

Ludwig, J., Meixner, F., Vogel, B., and Förstner, J.: Soil-air exchange of nitric oxide: An overview of processes, environmental factors, and modeling studies, Biogeochemistry, 52, 225-257, 2001.

Martin, R., and Asner, G.: Regional estimate of nitric oxide emissions following woody encroachment: linking imaging spectroscopy and field studies, Ecosystems, 8, 33-47, 2005.

Meixner, F. X.: Surface exchange of Odd Nitrogen Oxides, Nova Acta Leopoldina NF, 70, 299-348, 1994.

Meixner, F. X., Fickinger, T., Marufu, L., Serça, D., Nathaus, F. J., Makina, E., Mukurumbira, L., and Andreae, M. O.: Preliminary results on nitric oxide emission from a southern African savanna ecosystem, Nutrient Cycling in Agroecosystems, 48, 123-138, 1997.

Meixner, F. X., and Yang, W.: Biogenic emissions of nitric oxide and nitrous oxide from arid and semi-arid land, in: Dryland Ecohydrology, edited by: D'Odoricoand, P., and Porporat, A., Springer, Dordrecht, 233-255, 2006.

Michel, K., Matzner, E., Dignac, M.-F., and Kögel-Knabner, I.: Properties of dissolved organic matter related to soil organic matter quality and nitrogen additions in Norway spruce forest floors, Geoderma, 130, 250-264, 2006.

Millington, R. J.: Gas Diffusion in Porous Media, Science, 130, 100-102, 1959.

Millington, R. J., and Quirk, J. P.: Transport in porous media [in soil science], Trans International Congress Soil Sci 1960, 7, 97-106, 1960.

Moldrup, P., Olesen, T., Gamst, J., Schjonning, P., Yamaguchi, T., and Rolston, D. E.: Predicting the gas diffiusion coefficent in repacked soil: water-induced linear reduction model, Soil Science Society of America Journal, 64, 1588-1594, 2000.

Moravek, A.: Vertical distribution of reactive and non-reactive trace gases in and above a spruce canopy, MSc thesis, Department of Bauingeneur-, Geo- and Umweltwissenschaften, University of Karlsruhe, Karlsruhe, Germany, 126 pp., 2008.

Mosier, A. R., Mohanty, S. K., Bhadrachalam, A., and Chakravorti, S. P.: Evolution of dinitrogen and nitrous oxide from the soil to the atmosphere through rice plants, Biology and Fertility of Soils, 9, 61-67, 1990.

Muhr, J., Goldberg, S. D., Borken, W., and Gebauer, G.: Repeated drying-rewetting cycles and their effects on the emission of CO_2, N_2O, NO, and CH_4 in a forest soil, Journal of Plant Nutrition and Soil Science, 171, 719-728, 2008.

Nägele, W., and Conrad, R.: Influence of pH on the Release of NO and N_2O from Fertilized and Unfertilized Soil, Biology and Fertility of Soils, 10, 139-144, 1990a.

Nägele, W., and Conrad, R.: Influence of soil pH on the nitrate-reducing microbial populations and their potential to reduce nitrate to NO and N_2O, FEMS Microbiology Letters, 74, 49-57, 1990b.

Nishina, K., Takenaka, C., and Ishizuka, S.: Spatial variations in nitrous oxide and nitric oxide emission potential on a slope of Japanese cedar (*Cryptomeria japonica*) forest, Soil Science & Plant Nutrition, 55, 179-189, 2009.

Oberdorfer, E.: Pflanzensoziologische Exkursionsflora, Ulmer, Stuttgart, 1050 pp., 1994.

Olive, D. J.: Prediction intervals for regression models, Computational Statistics & Data Analysis, 51, 3115-3122, 2007.

Ormeci, B., Sanin, S. L., and Peirde, J. J.: Laboratory study of NO flux from agricultural soil: Effects of soil moisture, pH, and temperature, Journal of Geophysical Research, 104, 1621-1629, 1999.

Otter, L. B., Yang, W. X., Scholes, M. C., and Meixner, F. X.: Nitric oxide emissions from a southern African savanna, Journal of Geophysical Research-Atmospheres, 104, 18471-18485, 1999.

Paavolainen, L., and Smolander, A.: Nitrification and denitrification in soil from: A clear-cut Norway spruce (Picea abies) stand, Soil Biology & Biochemistry, 30, 775-781, 1998.

Papke, H., and Papen, H.: Influence of acid rain and liming on fluxes of NO and NO_2 from forest soil, Plant and Soil, 199, 131-139, 1998.

Philippot, L., Hallin, S., Borjesson, G., and Baggs, E. M.: Biochemical cycling in the rhizosphere having an impact on global change, Plant and Soil, 321, 61-81, 2009.

Pilegaard, K., Hummelshoj, P., and Jensen, N. O.: Nitric oxide emission from a Norway spruce forest floor, Journal of Geophysical Research-Atmospheres, 104, 3433-3445, 1999.

Pilegaard, K., Skiba, U., Ambus, P., Beier, C., Bruggemann, N., Butterbach-Bahl, K., Dick, J., Dorsey, J., Duyzer, J., Gallagher, M., Gasche, R., Horvath, L., Kitzler, B., Leip, A., Pihlatie, M. K., Rosenkranz, P., Seufert, G., Vesala, T., Westrate, H., and Zechmeister-Boltenstern, S.: Factors controlling regional differences in forest soil emission of nitrogen oxides (NO and N_2O), Biogeosciences, 3, 651-661, 2006.

Plake, D.: Vertikale Konzentrationsprofile und Flüsse von reaktiven und nicht reaktiven Spurengasen im Fichtelgebirge, MSc thesis, Landscapeecology, Westfälische Wilhelms-University Münster, Münster, Germany, 168 pp., 2009.

Remde, A., Slemr, F., and Conrad, R.: Microbial-production and uptake of nitric-oxide in soil, FEMS Microbiology Ecology, 62, 221-230, 1989.

Remde, A., Ludwig, J., Meixner, F. X., and Conrad, R.: A Study to Explain the Emission of Nitric-Oxide from a Marsh Soil, Journal of Atmospheric Chemistry, 17, 249-275, 1993.

Rudolph, J., and Conrad, R.: Flux between soil and atmosphere, vertical concentration profiles in soil, and turnover of nitric oxide: 2. Experiments with naturally layered soil cores, Journal of Atmospheric Chemistry, 23, 275-300, 1996.

Scheffer, F., and Schachtschabel, P.: Lehrbuch der Bodenkunde, Spektrum, Berlin, 593 pp., 2002.

Schindlbacher, A., Zechmeister-Boltenstern, S., and Butterbach-Bahl, K.: Effects of soil moisture and temperature on NO, NO_2, and N_2O emissions from European forest soils, Journal of Geophysical Research, 109, 10.1029/2004JD004590, 2004.

Schmitt, A., Glaser, B., Borken, W., and Matzner, E.: Repeated freeze-thaw cycles changed organic matter quality in a temperate forest soil, Journal of Plant Nutrition and Soil Science, 171, 707-718, 2008.

Skiba, U., Fowler, D., and Smith, K.: Emissions of NO and N_2O from soils, Environmental Monitoring and Assessment, 31, 153-158, 1994.

Skiba, U., Fowler, D., and Smith, K. A.: Nitric oxide emissions from agricultural soils in temperate and tropical climates: sources, controls and mitigation options, Nutrient Cycling in Agroecosystems, 48, 139-153, 1997.

Skopp, J., Jawson, M. D., and Doran, J. W.: Steady-state aerobic microbial activity as a function of soil water content, Soil Science Society of America Journal, 54, 1619-1625, 1990.

Slemr, F., and Seiler, W.: Field study of environmental variables controlling the NO emissions from soil and the NO compensation point, Journal of Geophysical Research, 96, 13,017-013,031, 1991.

Smith, K. A., Ball, T., Conen, F., Dobbie, K. E., Massheder, J., and Rey, A.: Exchange of greenhouse gases between soil and atmosphere: interactions of soil physical factors and biological processes, European Journal of Soil Science, 54, 779-791, 2003.

Smith, M. S., and Tiedje, J. M.: The effect of roots on soil denitrification, Soil Science Society of America Journal, 43, 951-955, 1979.

Stöhr, C., and Ullrich, W. R.: Generation and possible roles of NO in plant roots and their apoplastic space, Journal of Experimental Botany, 53, 2293-2303, 2002.

Stöhr, C., and Stremlau, S.: Formation and possible roles of nitric oxide in plant roots, Journal of Experimental Botany, 57, 463-470, 2006.

Subke, J.-A., Reichstein, M., and Tenhunen, J. D.: Explaining temporal variation in soil CO_2 efflux in a mature spruce forest in Southern Germany, Soil Biology and Biochemistry, 35, 1467-1483, 2003.

van Dijk, S. M., and Meixner, F. X.: Production and consumtion of NO in forest and pasture soils from the Amazon Basin, Water, air, and Soil Pollution, 1, 119-130, 2001.

van Dijk, S. M., Gut, A., Kirkman, G. A., Meixner, F. X., Andreae, M. O., and Gomes, B. M.: Biogenic NO emissions from forest and pasture soils: Relating laboratory studies to field measurements, Journal of Geophysical Research-Atmospheres, 107, 2002.

Venterea, R. T., and Rolston, D. E.: Mechanisms and kinetics of nitric nitrous oxide production during nitrification in agricultural soil, Global Change Biology, 6, 303-316, 2000.

Venterea, R. T., Groffman, P. M., Castro, M. S., Verchot, L. V., Fernandez, I. J., and Adams, M. B.: Soil emissions of nitric oxide in two forest watersheds subjected to elevated N inputs, Forest Ecology and Management, 196, 335-349, 2004.

Venterea, R. T., Rolston, D. E., and Cardon, Z. G.: Effects of soil moisture, physical, and chemical characteristics on abiotic nitric oxide production, Nutrient Cycling in Agroecosystems, 72, 27-40, 2005.

Vos, G., Bergevoet, I., Védy, J., and Neyroud, J.: The fate of spring applied fertilizer N during the autumn-winter period: comparison between winter-fallow and green manure cropped soil, Plant and Soil, 160, 201-213, 1994.

Wallenda, T., Stober, C., Högbom, L., Schinkel, H., George, E., Högberg, P., and Read, D. J.: Nitrogen uptake processes in roots and mycorrhizas, in: Carbon and nitrogen cycling in European forest ecosystems, edited by: Schulze, E. D., Springer, Heidelberg, 122-143, 2000.

Wardle, D. A., Bardgett, R. D., Klironomos, J. N., Setälä, H., van der Putten, W. H., and Wall, D. H.: Ecological linkages between aboveground and belowground biota, Science, 304, 1629-1633, 10.1126/science.1094875, 2004.

Wu, T., Sharda, J. N., and Koide, R. T.: Exploring interactions between saprotrophic microbes and ectomycorrhizal fungi using a protein-tannin comples as an N source by red pine (Pinus resinosa), New Phytologist, 159, 131-139, 2003.

Yu, J. B., Meixner, F. X., Sun, W. D., Liang, Z. W., Chen, Y., Mamtimin, B., Wang, G. P., and Sun, Z. G.: Biogenic nitric oxide emission from saline sodic soils in a semiarid region, northeastern China: A laboratory study, Journal of Geophysical Research, 113, 11, 2008.

Zheng, X. H., Huang, Y., Wang, Y. S., Wang, M. X., Jin, J. S., and Li, L. T.: Effects of soil temperature on nitric oxide emission from a typical Chinese rice-wheat rotation during the non-waterlogged period, Global Change Biology, 9, 601-611, 2003.

Appendix C

Distributed Modified Bowen Ratio Method for Surface Layer Fluxes of reactive and non-reactive Trace Gases

J.-C. Mayer[1], A. Bargsten[1], U. Rummel[2], F.X. Meixner[1,3] and T. Foken[4]

[1] Biogeochemistry Department, Max Planck Institute for Chemistry, Mainz, Germany

[2] Richard Aßmann Observatory Lindenberg, German Meteorological Service, Germany

[3] Department of Physics, University of Zimbabwe, Harare, Zimbabwe

[4] Department of Micrometeorology, University of Bayreuth, Bayreuth, Germany

Submitted to Agricultural and Forest Meteorology: March 2010

Accepted for Agricultural and Forest Meteorology: October 2010

Available online: 5 November 2010.

Doi:10.1016/j.agrformet.2010.10.001

Appendix C

Abstract

Modified Bowen ratio technique was used in a horizontally distributed form to determine turbulent fluxes of CO_2, H_2O, O_3, NO and NO_2 over a semi-natural grassland site in North-Eastern Germany. The applicability of the distributed variation of the modified Bowen ratio technique was proven prior to the calculation of trace gas fluxes. Turbulent NO fluxes were compared to fluxes up-scaled from laboratory measurements of biogenic NO emission from soil samples, which have been taken at the field site. The NO fluxes up-scaled from laboratory measurements were slightly larger than the fluxes observed in the field. However, both NO fluxes agreed within a factor of two. Under suitable night time conditions, we performed a detailed comparison of turbulent fluxes of CO_2 and O_3 with fluxes derived by the boundary layer budget technique. While there was agreement between these fluxes in a general sense, specific deviations were observed. They could be attributed to different footprint sizes of both methods and to in-situ chemistry within the nocturnal boundary layer.

Keywords

Flux; Trace gas; Exchange; Timescale; Turbulence

1 Introduction

Temporal and spatial variations of trace gas fluxes from and to the surface are crucial for understanding exchange processes between the atmosphere and terrestrial surfaces. A lot of effort in this respect has been done for the species carbon dioxide (CO_2), mostly using eddy covariance (EC) techniques (see e.g. Baldocchi et al., 1988; Suni et al., 2003; Baldocchi et al., 2001; Aubinet et al., 2000). Turbulent fluxes of reactive trace gases, e.g. ozone (O_3), nitrogen oxide (NO) and nitrogen dioxide (NO_2), are currently measured more on campaign than on a continuous basis (see e.g. Rummel et al., 2002; Wesely et al., 1982; Keronen et al., 2003). If fast enough sensors for EC measurements for these trace gases are available at all, they often need permanent maintenance, making them less suitable for permanent measuring networks.

An alternative technique to measure turbulent fluxes of trace gases is the so-called Modified Bowen Ratio (MBR) method as proposed by Businger (1986) and Müller et al. (1993). Originally, fluxes of trace gases have been derived from sensible heat flux (H) and vertical differences of air temperature (ΔT) and mixing ratio (Δc), where H has been typically determined by the Bowen Ratio method (Bowen, 1926; Lewis, 1995). With the development of sonic anemometers, a direct measurement of H became feasible, reducing the instrumental effort because radiation and soil heat flux measurements were no longer needed (Liu and Foken, 2001).

The traditional setup of a MBR station includes EC measurements of H and the determination of ΔT and Δc at the same location. Furthermore, ΔT and Δc have to be measured at the same heights. However, if several trace gases have to be measured and analyzer or inlet constructions are somewhat bulky, measurement errors due to flow distortion could be substantial. But this error source can be tackled (if a sufficiently large, homogeneous site is available) by using distributed locations for the individual measurements of Δc, ΔT and H. Moreover, for intercomparison studies, this reduces potential deviations between different measuring systems to the trace gas part of the measurements, as all flux calculations will relate on the same data set of H and ΔT. This approach will be presented in this paper and will be referred to as the Distributed Modified Bowen Ratio (DMBR) method.

The methods for measuring the fluxes of reactive trace gases are in principle the same as for non-reactive trace gases. The concept, whether the reactivity of a trace gas has to be taken into account was firstly described by Damköhler (1940). Therefore, the ratio of the characteristic time scale for chemical reactions to the time scale of turbulence is referred to as the Damköhler number (DA). As long as DA is much smaller than one, the reactivity can be neglected. Otherwise, chemical alteration during the transport must be considered. A determination of individual fluxes of reactive trace gases independent from the risk of chemical alteration would be a great advantage. In the case of NO, soil emission fluxes, derived from laboratory measurements on soil samples, would be a suitable approach. However, laboratory measurements do provide only a sound parameterization of soil NO fluxes under varying soil moisture and soil temperature conditions; hence, NO fluxes up-scaled from laboratory measurements will never have the state of actual field measurements, which are made under variable ambient conditions. The comparison of laboratory derived fluxes with actual field fluxes will always address this principal difference between laboratory and field methods.

A similar statement holds, if flux measuring methods differ in the characteristic transport time relevant to the measurements which are being used to compute the fluxes. In case, the trace gas flux is derived from the vertical mixing ratio difference, the characteristic transport time increases, if the vertical spacing between two levels is increased. This, in turn, increases the potential influence of chemical reactions to change the mixing ratios of reactive trace gases (during turbulent transport). This has to be taken into account, when comparing fluxes obtained by methods with large to those with only small spatial extent.

If vertical profile data of a quantity, whose flux has to be determined, are available up to the equilibrium height (i.e. were temporal changes of mixing ratios are not observable anymore), the so-called boundary layer budget method can be used (Pattey et al., 2002). This method integrates vertical profiles of the desired quantity and assigns its temporal change to a vertical flux into the corresponding volume (Denmead et al., 1996; Levy et al., 1999). If horizontal advection can be excluded, and the top end of the profile is capped by a "lid", the temporal change should equal to the vertical flux at the bottom end of the profile, i.e. the flux determined by DMBR or EC methods. The "lid" can either be a strong inversion, or the presence of strong wind shear due to a low-level jet (Mathieu et al., 2005). As a case study, we compare fluxes of sensible heat, CO_2 and O_3 obtained from the DMBR method to the respective fluxes obtained from the boundary layer budget method. This comparison should show (a) the similarity of both methods for conservative quantities and (b) the increasing influence of chemical reactions (on O_3 flux) for the method with the larger spatial extension (the boundary layer budget method).

Appendix C

2 Material and Methods

2.1 Site and Setup

The LIBRETTO (LIndenBerg REacTive Trace gas prOfiles) campaign took place in late summer 2006, from 01 August 2006 until 31 August 2006 at the Falkenberg Boundary Layer Field Site of the Meteorological Observatory Lindenberg (Richard-Aßmann Observatory) (Beyrich and Adam, 2007). The field site is located at 52° 10' 01" N, 14° 07' 27" E, 73 m a.s.l. The main vegetation species are perennial ryegrass (*Lolium perenne*), red fescue (*Festuca rubra*), dandelion (*Leontodon autumnalis*, *Taraxacum officinale*), bromegrass (*Bromus hordeaceus*), and clover (*Trifolium pratense*, *Trifolium repens*). The meadow is mowed regularly in order to keep the mean vegetation height below 20 cm (Beyrich and Adam, 2007). However, during the LIBRETTO campaign, the vegetation height was between 5 cm and 8 cm. The measuring site comprises one 99 m and one 10 m high profile mast (air temperature (T), relative humidity (rH), wind speed (u) and wind direction), two identical setups for the measurement of the net radiation flux, two stations for the measurement of turbulent fluxes of momentum, sensible and latent heat (further on referred to as the EC stations), and a sub-site to monitor physical soil quantities (soil temperatures, soil heat flux and soil moisture). A SODAR-RASS system completes the permanent setup of the Falkenberg site. Details about the instrumentation relevant for this work are summarized in Table 1, the spatial situation is shown in Figure 1. All heights given in this study are heights above ground level, unless otherwise stated.

Appendix C

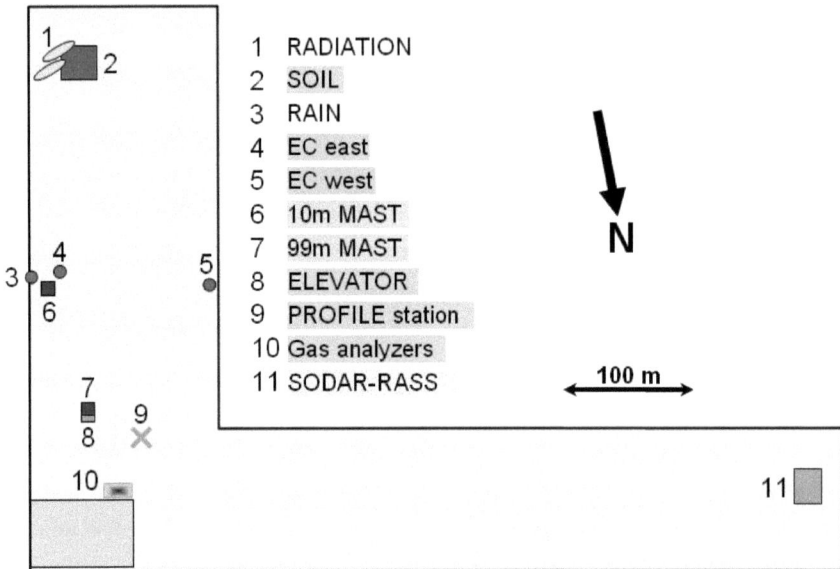

Figure 1: Spatial arrangement of the different stations at the Falkenberg Boundary Layer Field Site. Additional instrumentation ("Profile station") during the LIBRETTO campaign is indicated by orange color.

The 99 m profile mast is equipped with an elevator, usually used for service and maintenance. On this elevator, additional sensors for T and rH as well as for ozone (O_3), carbon dioxide (CO_2) and water vapour (H_2O) have been installed (see Table 1). However, we limit our evaluation from this system to T, CO_2 and O_3, because the H_2O channel of the deployed instrument has unfortunately not worked properly. The elevator was automatically run up and down once every 10 minutes, interrupted only for data retrieving and service. At bottom and top position, the elevator had approx. 6.5 minutes idle time for equilibration intercomparison with stationary sensors and analyzers, thus one profile needed approx. 3.5 minutes to be measured. A detailed description of the algorithms which have been applied to correct the dynamical error of the elevator based measurements are given by Mayer et al. (2009).

For the LIBRETTO campaign, an additional set of profile instruments was installed comprising the measurement of the trace gases CO_2, H_2O, O_3, NO and NO_2 at three levels and air temperature at 4 levels (0.25, 0.5, 1.0, 2.0 m). The trace gas inlets were located at 0.15 m, 1.0 m and 2.0 m. A fourth sampling tube collected air from the base position of the elevator (at 2.0 m) for continuous adjustment of the ground based mixing ratio measurements and those measured on the elevator. Air samples were pumped from the corresponding intake devices (i.e. downward directed funnels and particle filters) via heated Teflon tubes to a switching valve manifold (see Figure 2), located next to the trace gas analyzers, housed in an air conditioned container, approx. 50 m NE from the "PROFILE station" (see Figure 1). After completion of set-up and test of the instrumentation, the measuring period of the LIBRETTO campaign consisted of 20 days, from 11 to 30 August, 2006.

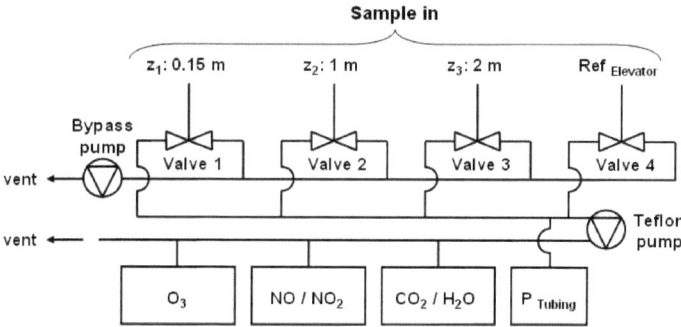

Figure 2: Scheme of the gas flow of the switched trace gas profile. A customized control system (PC based) was used for (a) switching valves 1 – 4, (b) data acquisition, and (c) controlling the NO / NO_2 analyzer (to be in phase with valve switching). While only one sample line per time was connected with the sampling Teflon pump, the three other sampling lines were flushed with a bypass pump to avoid stagnant air and increased lag times.

Appendix C

Table 1. Setup of the field site during the LIBRETTO campaign. Only instruments being relevant for this work are listed.

Parameter	Symbol	Unit	Temporal resolution (min)	Sampling height (m) a.g.l.	Instrument (Model)
Permanent setup					
Sensible Heat Flux	H	Wm^{-2}	10	2.4	METEK USA1
Air temperature, EC station West/East	T	°C	10	0.55/0.5, 2.8/2.4	Pt-100
Friction velocity	u_*	ms^{-1}	10	2.4	METEK USA1
Global radiation	Rg	Wm^{-2}	10	2.4	Kipp&Zonen CM22
Air temperature, 10m profile	T	°C	10	0.5, 4	Pt-100
Wind speed	U	ms^{-1}	10	0.5, 4	Climatronics F460
Rain	R	mm	10	1	Ott, Pluvio
Additional LIBRETTO setup					
Ground based					
Air temperature	T	°C	10	0.25, 0.5, 1.0, 2.0	Aspirated Thermocouple
Carbon dioxide	CO_2	ppm	10	0.15, 2.0	LiCor LI 7000
Water vapour	H_2O	‰	10	0.15, 2.0	LiCor LI 7000
Nitrogen monoxide	NO	ppb	10	0.15, 2.0	EcoPhysics CLD 780

| Nitrogen dioxide | NO$_2$ | ppb | 10 | 0.15, 2.0 | EcoPhysics CLD 780 |
| Ozone | O$_3$ | ppb | 10 | 0.15, 2.0 | Thermo Electron 49C |

Additional LIBRETTO setup
Elevator

Air temperature	T	°C	10	2 - 99	Thermocouple
Carbon dioxide	CO$_2$	ppm	10	2 - 99	LiCor LI 840
Water vapour	H$_2$O	‰	10	2 - 99	LiCor LI 840
Air pressure	P	hPa	10	2 - 99	Vaisala PTB 101B
Ozone	O$_3$	ppb	10	2 - 99	GFAS OS-G-2

2.2 Quality Control and gap filling

2.2.1 Reference data

Routinely measured data from the permanent setup of the Lindenberg site are called "reference data" in the following, in order to contrast the experimental data from those instruments which have been added to the site during LIBRETTO (see Table 1).

Reference data (from EC and radiation flux stations, data of soil physical quantities) were flagged with quality indicators according to Beyrich and Adam (2007). Because the EC stations were located at the western and eastern side of the south-leg of the field site (see Figure 1), at least one EC dataset (being representative for the field site) was available under every "fetch" (upwind) conditions. A so-called "limited fetch condition" occurred for the western EC station at wind directions from 180°-360° and for the eastern EC station from

Appendix C

30°-150°. These data are flagged correspondingly. Only data without any critical flag passed to further analyses.

Friction velocity (u*) and sensible heat flux (H) are key parameters in most of our evaluations. Therefore, special attention was paid to gap-filling procedures for these parameters. If EC data were not available (critical flag), u* and H were computed according to Arya (2001) from ΔT and Δu, measured at z = 0.5 m and z = 4.0 m at the 10 m profile mast. We made use of the following equations:

(a) geometrical mean height z_m of the layer (z_1; z_2):

$$z_m = \sqrt{z_1 \cdot z_2} \ . \tag{1}$$

(b) Richardson number Ri (a measure of dynamic stability of the layer (z_1; z_2)) as function of ΔT and Δu:

$$Ri = \frac{g}{T_1} \cdot \frac{\Delta\Theta \cdot z_m}{(\Delta u)^2} \ln \frac{z_2}{z_1} \ , \tag{2}$$

where g stands for the acceleration due to gravity, T1 is the absolute temperature at level z, and Θ denotes the potential temperature. The following relation exists between Ri and the Monin-Obukhov stability parameter ζ:

$$\zeta = Ri \qquad \text{if } Ri < 0 \text{ (unstable)}$$

$$\zeta = \frac{Ri}{(1 - 5 \cdot Ri)} \qquad \text{if } 0 \leq Ri < 0.2 \text{ (neutral to stable)} \tag{3}$$

(c) u* and H:

$$u_* = \frac{\kappa \cdot \Delta u}{\left(\varphi_m \cdot \ln \frac{z_2}{z_1} \right)} \tag{4}$$

$$T_* = \frac{\kappa \cdot \Delta T}{\left(\varphi_H \cdot \ln \frac{z_2}{z_1} \right)} \ . \tag{5}$$

Appendix C

$$H = -T_* \cdot u_* \cdot \rho_{air} \cdot c_p, \qquad (6)$$

where c_p is the heat capacity of air at constant pressure and ρair denotes the density of air. κ is the von Kármán constant (0.4). For the universal functions φ_m and φ_H we make use of those given by Businger et al (1971) with the modifications by Högström (1988).

Before and after each gap, the deviation of the profile-derived values of u_* and H were adjusted against the EC values. If a trend was present, it was distributed linearly over the gap. A total of seven 30 minute data points out of 1056 data points for H and u* were filled with this method for the entire experiment.

2.2.2 Profile station data

Two primary data sets were obtained from the profile station: (a) air temperature data and (b) trace gas data. Data recorded during servicing the profile station, farming activities (within the fetch), or rainfall periods were replaced by a wildcard. Subsequently, calibration factors obtained from calibration activities prior, during and after the experiment were applied. Identification of spikes was performed by a de-spiking scheme based on that of Vickers and Mahrt (1997). If more than 3 consecutive spikes were detected, they were treated as valid data – all other spikes were replaced by wild cards. Next, a gap filling procedure based on non-linear interpolation (Akima, 1970) was applied, but limited to a maximum gap size of 30 minutes. As a last step, all data were averaged to an interval of 30 minutes for further evaluation.

2.3 Distributed Modified Bowen Ratio (DMBR)

The MBR as well as the DMBR method requires (a) simultaneous measurements of ΔT and Δc at identical levels and (b) that the measured H is representative for the same source area (i.e. that area which is influencing the flux measurements). Due to technical reasons, the lower level (z_1) for trace gas measurements was 0.15 m, while it was 0.25 m for the temperature measurements. The upper level (z_2) was 2.0 m in both cases. Therefore, a ΔT_c corresponding to a height interval from 0.15 m to 2.0 m had to be computed from the

Appendix C

measured ΔT_m between 0.25 m and 2.0 m by means of the Monin-Obukhov similarity theory. For that, the state of atmospheric stability can be expressed by the Obukhov length L:

$$L = -\frac{u_*^3}{\kappa \dfrac{g}{T} \dfrac{H}{c_p \cdot \rho_{air}}}. \tag{7}$$

The calculation of L was performed using H and u* from the EC measurements. Within the surface layer, fluxes are assumed to be constant with height, thus L is also independent of z. Then, ΔT_c can be computed from ΔT_m by:

$$\Delta T_c = \Delta T_m \times \frac{\left[\ln\dfrac{z_{2,c}}{z_{1,c}} - \Psi_H\left(\dfrac{z_{2,c}}{L}\right) + \Psi_H\left(\dfrac{z_{1,c}}{L}\right)\right]}{\left[\ln\dfrac{z_{2,m}}{z_{1,m}} - \Psi_H\left(\dfrac{z_{2,m}}{L}\right) + \Psi_H\left(\dfrac{z_{1,m}}{L}\right)\right]}. \tag{8}$$

The index m refers to the measured values and index c refers to computed values. Ψ_H denotes the integrated universal stability function for H (Businger et al., 1971) with the modifications according to Högström (1988). For unstable stratification, this integration in general form was firstly done by Paulson (1970). Due to the low vegetation height (see Section 2.1), the displacement height was neglected.

The calculation of ΔT_c (at "PROFILE station") requires the sensible heat flux (H), which was measured at both EC stations (see Section 2.2.1), which were located about 150 m SSW and 165 m SSE of the "PROFILE station" (see Figure 1). This spatial separation of measurements of H and ΔT (and Δc) is characteristic of the DMBR method, and it is obvious, that the successful application of the DMBR demands horizontal homogeneity of the measurement site. A first-order check of this condition could be provided. Simultaneously to the EC measurements of H, measurements of air temperature were performed at $z = 0.55$ m and $z = 2.8$ m (west EC station) and at $z = 0.5$ m and $z = 2.4$ m (east EC station). By applying Eq. 8 to this measured temperature difference, it was recalculated to match the measurement levels of T at the "PROFILE station" (0.5 m and 2.0 m). By the comparison of ΔT measured at the "PROFILE station" and ΔT computed at the EC station, a good agreement would indicated the fulfillment of horizontal homogeneity with respect to sensible heat at the site.

Finally, under the assumption of identical eddy diffusivities for the scalars sensible heat and trace gases, fluxes ($F_{TraceGas}$) of the trace gases CO_2, H_2O, O_3, NO and NO_2 were computed from observed H, Δc and ΔT_c according to Foken (2008):

$$F_{TraceGas} = \frac{\Delta c}{\Delta T_c} \cdot H. \tag{9}$$

Obtained flux values were discarded for quality reasons, if $u_* < 0.07$ m s^{-1} (Liu and Foken, 2001), $\Delta T_c < 0.2$ K or the stability parameter ζ was outside ± 1, indicating either very strong stability or free convection.

2.4 Laboratory measurements

Net NO release rates were determined using an automated dynamic chamber laboratory system. A detailed description of our experimental setup is given in van Dijk and Meixner (2001). We present here only a brief description of the set-up.

Pressurized NO-free air supplied five Plexiglas cuvettes (four incubation cuvettes (each containing 100 g of soil samples) and one empty reference cuvette) with a continuous flow of 4.17 m^3 s^{-1} (2.5 l min^{-1}) per cuvette. All connections and tubes consisted of inert material (PTFE). The outlet of each cuvette was connected to a switching valve. Every two minutes one cuvette was connected to a chemiluminescence NO-analyzer (Model 42i Trace Level, Thermo Electron Corporation, United States, detection limit at 3σ: 250 ppt) and a CO_2/H_2O-analyzer (LI-840, Licor, United States), while the remaining four cuvettes were still purged, so that all five cuvettes were measured within 10 minutes. The NO-analyzer was calibrated using a gas phase titration unit (GPT, 146 C Dynamic Gas Calibrator, Thermo Electron Corporation, United States) supplied with NO from a NO standard cylinder (5.02 ppm NO, Air Liquide, Germany).

The loss of water (measured in terms of water vapour) was related to the gravimetric soil moisture content at the start and the end of the laboratory analyses. The gravimetric soil moisture content was converted into units of water filled pore space (WFPS) using the field

Appendix C

bulk density (1.245 g cm^{-3}) and the particle density of the average soil material (2.65 g cm^{-3}). This is particularly useful, as the WFPS is not only a term characterizing the soil water available to microorganisms, but represents also the soil air-water ratio, which controls the gaseous diffusion into or out of the soil (Ormeci et al., 1999).

Mixed soil samples were taken in May 2008 on the Lindenberg site. 25 samples, randomly taken from an area of approx 1000 m^2 within the fetch, yielded a total of approx. 3 kg of soil, from which subsamples for 3 repetitive analyses were separated. They were air-dried, sieved (< 2 mm) and stored at 4 °C. All samples were measured within 4 weeks after sampling. The NO measurements were conducted at three different temperatures (10, 20 and 30 °C), two different NO mixing ratios (0 and 50 ppb NO) and over a full range of soil moisture (0 - 100 % WFPS). Twelve hours before starting the laboratory measurement the soil samples were preincubated: the soil samples were moisturized with deionised water to about 100 % WFPS and stored in a temperature controlled cabinet at the temperature of subsequent laboratory measurements.

The net NO release rate (J) was calculated from the mixing ratio difference between the reference cuvette (NO$_{ref}$) and the incubation cuvettes (NO$_{out}$):

$$J = \left(\frac{Q}{M_{soil}}\right) \times \left([NO_{ref}] - [NO_{out}]\right) \times \left(\frac{M_N}{V_m}\right) \qquad (10)$$

where Q is the flow through the cuvette (m^3 s^{-1}), M$_{soil}$ is the sample weight (kg), M$_N$ is the molecular weight of nitrogen (14.0076 kg kmol^{-1}) and V$_m$ is the molar volume (m^3 kmol^{-1}) at actual temperature and standard pressure (1013.25 hPa). The results from all 3 repetitive measurements performed at one temperature were combined to calculate the corresponding NO release rate.

From the obtained NO release rates, a soil temperature and soil moisture dependent net potential NO flux (in units of ng m^{-2} s^{-1}) was derived, applying the Galbally & Johansson algorithm (1989), which has been modified to account for the dependence of the net potential NO flux on soil temperature and soil moisture (Feig et al., 2008; Yu et al., 2008; Bargsten et al., 2010). For comparison of laboratory derived NO fluxes with those measured by the DMBR method (see Section 3.5.1), net potential NO fluxes have been up-scaled by actual field data of soil temperature and soil moisture, which were measured during the LIBRETTO campaign.

2.5 Boundary layer budget method

The boundary layer budget method is based on calculating the budget of any quantity within the boundary layer (Denmead et al., 1996; Eugster and Siegrist, 2000). A surface flux of trace gases into (emission) or out of (deposition) the boundary layer changes the amount of the considered trace gas within the boundary layer. Vertical integration of the mixing ratio yields the total mixing ratio being present at a certain time. The upper boundary of integration must be at least the height where no temporal change of mixing ratio due to surface fluxes occurs. The change of mixing ration between two subsequent integrations thus should equal the surface flux during this period. This approach however is only valid in the absence of advection. The upper integration limit is usually set to the top of the nocturnal boundary layer (NBL) at nighttime or to the top of the convective boundary layer (CBL) during day.

2.6 Characteristic time scales

The characteristic time scales denote the time needed for certain processes to occur. In case of the turbulent time scale τ_{turb}, it is a measure of the intensity of turbulent transport. For near neutral conditions, τ_{turb} can be expressed as

$$\tau_{turb} = \frac{\kappa \cdot (z + z_0)}{A \cdot u_*} , \tag{11}$$

with the roughness length z_0 and the height above ground z (Vilà-Guerau de Arellano and Duynkerke, 1992). A stands for the ratio of the vertical velocity variance and the friction velocity squared. However, this formulation is not suitable to describe diel cycles, because this formulation is only valid for near neutral conditions (Wyngaard, 1982). Furthermore, under neutral conditions, A is constant. As an alternative, τ_{turb} can be computed from the turbulent diffusion coefficient (K) and the height (h) of the layer being considered (Rohrer et al., 1998):

Appendix C

$$\tau_{turb} = \frac{h^2}{K} . \tag{12}$$

Because of the linear dependence of K on height, the turbulent diffusion coefficient for momentum (K_m) can be expressed as

$$K_m = \kappa \cdot u_* \cdot z , \tag{13}$$

and the mean K_m across the considered layer as

$$K_m = \kappa \cdot u_* \cdot \frac{h}{2} . \tag{14}$$

For a layer from the ground up to h, τ_{turb} thus reads as

$$\tau_{turb} = \frac{2 \cdot h}{\kappa \cdot u_*} . \tag{15}$$

If the layer under consideration does not extend down to the ground and has the thickness Δz, the above equations must be modified. The mean K_m for the layer thus reads as

$$K_m = \kappa \cdot u_* \cdot \left(h - \frac{\Delta z}{2} \right) \tag{16}$$

and τ_{turb} can then be expressed as

$$\tau_{turb} = \frac{\Delta z}{\kappa \cdot u_*} \cdot \left[\frac{h}{\Delta z} - \frac{1}{2} \right]^{-1} . \tag{17}$$

To account for the influence of atmospheric stability on τ_{turb}, Eq. (13) can be modified by introducing the universal function φ_m:

$$K_m = \frac{\kappa \cdot u_* \cdot z}{\varphi_m} , \tag{18}$$

and τ_{turb} is then expressed as

$$\tau_{turb} = \frac{\varphi_m \cdot \Delta z}{\kappa \cdot u_*} \cdot \left[\frac{h}{\Delta z} - \frac{1}{2} \right]^{-1} . \tag{19}$$

So far, τ_{turb} yields the turbulent time scale for the transfer of momentum through a layer Δz with its upper boundary being h. Because of the non-similar intensity of the transfer of momentum and the transfer of heat or matter, the formulation of τ_{turb} for the latter quantities must include the turbulent Prandtl number Pr_t or turbulent Schmidt number Sc_t, respectively. They are defined as the ratio of the turbulent diffusion coefficient for momentum to that of heat (K_H) or the respective trace gas (K_C):

$$Pr_t = \frac{K_m}{K_H}, \qquad (20)$$

$$Sc_t = \frac{K_m}{K_C}, \qquad (21)$$

For heat and matter, the characteristic turbulent time scale thus reads as

$$\tau_{turb,HEAT} = \frac{\varphi_m \cdot \Delta z \cdot Pr_t}{\kappa \cdot u_*} \cdot \left[\frac{h}{\Delta z} - \frac{1}{2}\right]^{-1} \qquad (22)$$

$$\tau_{turb,MATTER} = \frac{\varphi_m \cdot \Delta z \cdot Sc_t}{\kappa \cdot u_*} \cdot \left[\frac{h}{\Delta z} - \frac{1}{2}\right]^{-1}. \qquad (23)$$

According to Foken (2008), Pr_t and Sc_t are both assumed to be 0.8.

For the chemical time scales τ_{chem}, we consider only the triad of NO-NO$_2$-O$_3$ including the (photo-)chemical reactions

$$O_3 + NO \xrightarrow{k_1} NO_2 + O_2 \quad k_1 = 1.40 \cdot 10^{-12} \exp\left(\frac{-1310}{T}\right) \qquad (R1)$$

where k_1 is taken from Atkinson et al (Atkinson et al., 2004), and

$$NO_2 + O_2 \xrightarrow{h\nu} NO + O_3 \quad h\nu < 420 \text{ nm.} \qquad (R2)$$

Confined to the NO-NO$_2$-O$_3$ triad, the individual time scale of NO$_2$ (τ_{NO2}) depends only on the actual NO$_2$ photolysis radiation flux, giving jNO$_2$. According to Trebs et al. (2009), there is a second order polynomial relationship between global radiation (R_g, in Wm^{-2}) and jNO$_2$ (in s^{-1}):

Appendix C

$$jNO_2 = B_1 \cdot R_g + B_2 \cdot R_g^2 \qquad (24)$$

where $B_1 = 1.47 \cdot 10^{-5}$ W^{-1} m^2 s^{-1} and $B_2 = -4.94 \cdot 10^{-9}$ W^{-2} m^4 s^{-1}.

When only the triad NO-NO$_2$-O$_3$ is considered, the photolysis frequency of NO$_2$ controls the destruction of NO$_2$ and thus reformation of NO and O$_3$. Under night time conditions, NO$_2$ is assumed to have a constant time scale of about 2 days (for calculation purposes only, namely to avoid an infinite time scale due to absence of any shortwave radiation). The individual time scale of O$_3$ and NO depends on the reaction rate k_1 for (R1) and the NO number density (N$_{NO}$ in molecules cm^{-3}) in case of τ_{O3}, and on the O$_3$ number density (N$_{O3}$) in the case of τ_{NO}:

$$\tau_{O_3} = \frac{1}{k_1 \cdot N_{NO}} \qquad (25)$$

$$\tau_{NO} = \frac{1}{k_1 \cdot N_{O_3}} \qquad (26)$$

$$\tau_{NO2} = \frac{1}{jNO_2} \qquad (27)$$

The overall chemical time scale of the NO-NO$_2$-O$_3$ triad, τ_{chem}, is given by Lenschow (1982). It provides a measure about the intensity of the chemical conversions for comparison with the turbulent time scale (τ_{turb}):

$$\tau_{chem} = \frac{2}{\sqrt{j_{NO_2} + k_1^2 \cdot (N_{O_3} - N_{NO})^2 + 2 \cdot k_1 j_{NO_2} (N_{O_3} + N_{NO} + 2 \cdot N_{NO_2})}} \qquad (28)$$

Given τ_{turb} and τ_{chem}, the Damköhler number (DA) is defined as:

$$DA = \frac{\tau_{turb}}{\tau_{chem}} \qquad (29)$$

DA << 1 indicates, that turbulent transport is much faster than (photo-)chemical reactions, which would allow reactive species to be treated as quasi passive. DA ≤ 1 indicates, that (photo-)chemical reactions during turbulent transport have to be taken into account to correctly derive turbulent fluxes of reactive species. An overview about typical magnitudes for τ_{turb} and τ_{chem} is given by Foken et al. (1995).

3 Results and Discussion

The results chapter is divided into five parts. They cover (a) the aspect of horizontal homogeneity, (b) characteristic time scales, (c) an overview about the atmospheric conditions for turbulent exchange, (d) median diel cycles of trace gas mixing ratios, and (e) median diel cycles of trace gas fluxes achieved by the DMBR method. The last part additionally comprises the comparison between (1) NO fluxes obtained by field measurements (DMBR) and those up-scaled from laboratory derived net potential NO fluxes and (2) between night time fluxes (H, CO_2, O_3) obtained by the DMBR method and by the nocturnal boundary layer budget method.

3.1 Horizontal homogeneity

Temperature differences at the EC stations (0.55 m / 2.8 m and 0.5 m / 2.4 m, respectively) were recalculated to match the measuring heights of the profile station (0.5 m and 2.0 m) by utilizing the MO theory (see Eq. (8)). Figure 3 shows the comparison between the computed ΔT at the EC station and the actually measured ΔT at the profile station, which resulted in a very significant correlation ($R^2 = 0.94$, n = 1007). This indicates the fulfillment of the precondition of horizontal homogeneity at the Lindenberg experimental site.

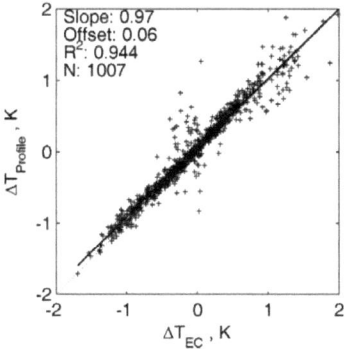

Figure 3: Comparison of measured temperature differences at the profile station with the computed temperature differences at the EC station. The dashed lines give the 1:1 ratio, the solid grey lines indicate the linear regression.

3.2 Characteristic time scales

As soon as the (turbulent) transport of reactive trace gases is investigated, their characteristic chemical time scales (τ_{chem}) must be considered and compared to the characteristic turbulent time scale (τ_{turb}). If τ_{turb} is considerably smaller than τ_{chem} (DA $\ll$ 1), i.e. transport is much faster than chemical reactions during the transport, the trace gases could be treated like non-reactive trace gases. In case of τ_{chem} is close to τ_{turb} (DA $\leq$ 1), chemical reactions during the transport have to be taken into account to correctly derive turbulent fluxes of the reactive gases. For the entire LIBRETTO campaign, τ_{turb} and τ_{chem} for the NO-NO$_2$-O$_3$ triad is shown in Figure 4a. During daytime, τ_{turb} was always smaller than τ_{chem}, demonstrating that turbulent transport was always faster than chemical interconversion between NO, NO$_2$ and O$_3$. Lowest values of τ_{turb} and τ_{chem} were observed around noon and in the early afternoon. These low values coincided with highest turbulence intensity and fastest chemistry (highest photolysis rate of NO$_2$). During night time, where NO$_2$ photolysis is excluded, the life time of NO$_2$ is infinite with respect to the NO-NO$_2$-O$_3$ triad. Thus τ_{chem} is fully controlled by reaction R1. However, the shapes of diel variations of τ_{turb} and τ_{turb} were similar, with τ_{turb} being typically 10–20 % of τ_{chem}. This is demonstrated by median diel cycle of the ratio of τ_{chem} to τ_{turb}, the Damköhler Number DA (Figure 4b). Maximum values of DA were found to occur during the stabilization of the atmospheric boundary layer in the late afternoon and during the first part of the night, while minimum values of DA where observed in the hours before sunrise.

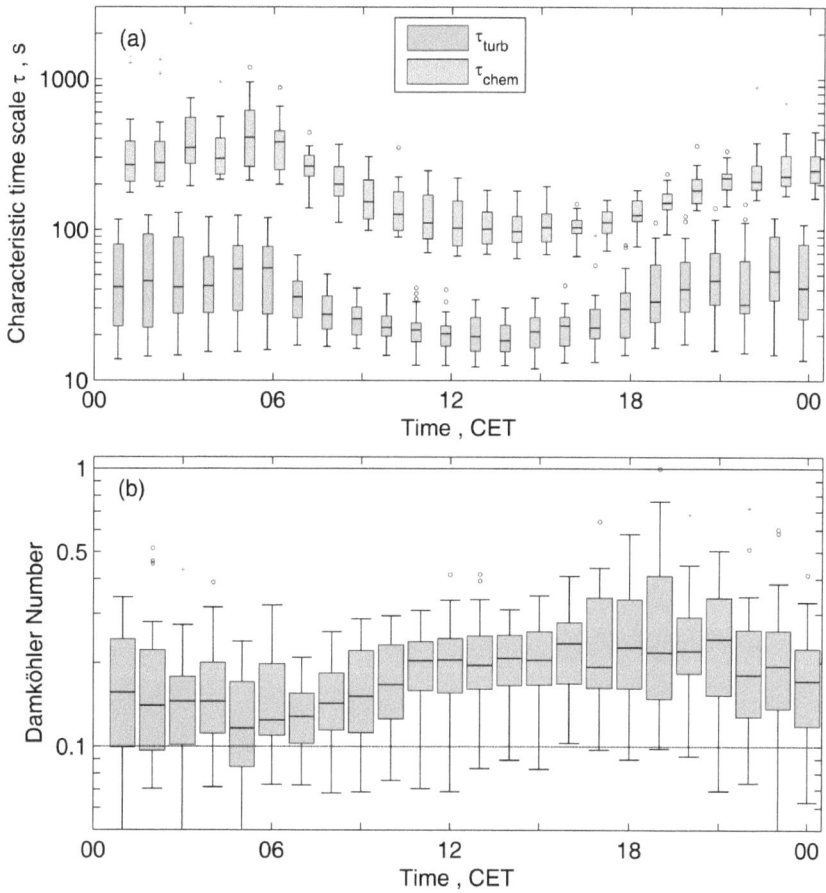

Figure 4: Median diel cycles for the period of 11 August 2006 – 30 August 2006: (a) characteristic time scales of turbulent transport (τ_{turb}) and (photo-)chemical reactions ($\tau_{chem.}$) for the layer 0.15–2.0 m a.g.l., (b) dimensionless Damköhler number (DA) for the NO-NO_2-O_3 triad. The black solid lines indicate the medians, the boxes cover the inter-quartile range. Blue circles denote extreme values, red crosses denote outliers.

Appendix C

The time of lowest DA values, which mean lowest influence of (photo-)chemical reactions on the trace gas mixing ratios and fluxes, coincided with the time of minimum O_3 mixing ratios (c.f. Figure 6e). Low O_3 mixing ratios lead to low NO destruction (see reaction R1). Because nighttime photolysis of NO_2 does not exist, reaction R2 did not occur during night. Therefore, even with low turbulence (high values of τ_{turb}), the impact of chemical reactions on the the nighttime observations was minimal. During daytime, when photolysis of NO_2 occurs and higher O_3 mixing ratios are present, the influence of (photo-)chemical reactions was slightly higher than before sunrise, but remained moderate (due to fast turbulent transport). The observed range of DA (0.1–0.2) indicates, that influences due to (photo-)chemical reactions were weak enough to treat the reactive trace gases as quasi passive in our further evaluations.

3.3 Thermodynamic conditions of exchange

Because trace gases interact with the ground surface, their mixing ratios within the surface layer are strongly dependent on exchange conditions and thus on atmospheric stability. Figure 5a shows the median diel course of u_*, Figure 5b shows that of the stability parameter ζ, determined at $z = 2.4$ m (i.e. the height of the EC measurements). During night (18:00 h – 06:00 h), low values of u_* around 0.1 m s^{-1} indicated prevalence of only low turbulence, allowing a possible accumulation of soil emitted trace gases (e.g. NO, CO_2) or a depletion of those trace gases having a net sink at the ground surface (NO_2, O_3). This is also supported by ζ, showing positive values during night (i.e. stable stratification). Around 06:00 h, stability changed to unstable conditions. Coincidentally, u_* increased and reached values around 0.3 m s^{-1} at noon. Strongest unstable conditions, however, were already observed around 09:00 h. When the atmosphere became stable again around 18:00 h, u_* decreased rapidly to low nocturnal values.

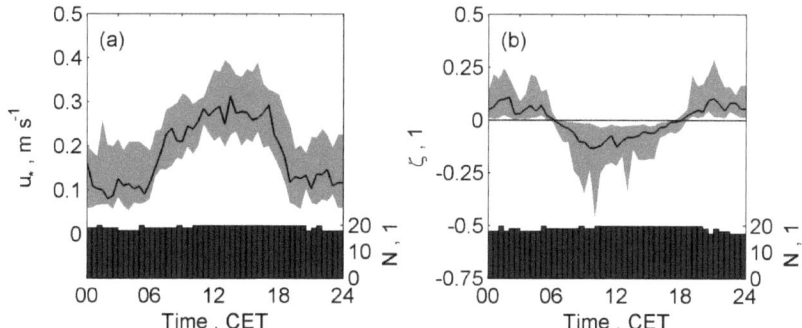

Figure 5: Median (straight lines) diel courses of (a) friction velocity u* and (b) atmospheric stability ζ, 11–30 August, 2006. Shaded areas represent respective inter-quartile ranges. Blue bars at the bottom of each graph indicate the number of data points available for calculation of medians and inter-quartile ranges.

3.4 Trace gases - mixing ratios

The median diel courses of all measured scalars (CO_2, H_2O, O_3, NO, NO_2, T) from 11 August 2006 to 30 August 2006 are shown in Figure 6 (left panels) together with their vertical mixing ratio differences (Δ) between 0.15 m and 2.0 m (right panels). The temporal resolution was 30 minutes. Highest values of CO_2 mixing ratio (≈ 420 ppm, Figure 6a) were reached in the second part of the night. At this time, ΔCO_2 (Figure 6b) was about -40 ppm, indicating strong accumulation processes close to the ground surface, where microbial processes in soil and plants constitute the source of CO_2. After sunrise, CO_2 gradually decreased until minimum mixing ratios of 365 ppm are reached around 12:00 h. ΔCO_2 changed its sign to positive values at 07:00 h, indicating a net CO_2 sink at the ground (photosynthesis). However, the magnitude of ΔCO_2 remained very small during the day, being lower than $+1$ ppm. In the afternoon, CO_2 mixing ratios started to increase again, and also ΔCO_2 became negative again (after 17:00) and larger. The water vapour (Figure 6c, 6d) mixing ratio exhibited a period of slight increase from 14 ppth to 15 ppth between 06:00 h and 09:00 h followed by a slight decrease until 12:30 h. During the afternoon, H_2O remained around 14 ppth until it changed around 18:00 h to about 15 ppth. During the night, the H_2O mixing ratio gradually decreased

Appendix C

to about 14 ppth at the end of the night. The H_2O mixing ratio difference, ΔH_2O, showed a much clearer diel cycle. At 06:00 h, it became negative, indicating evapotranspiration. Between 08:00 h and 15:00 h, ΔH_2O remained around -1.5 ppth. Afterwards, it slowly increased and became positive to the end of the night. The median diel course of O_3 (Figure 6e, 6f) showed minimum mixing ratios of about 12 ppb in the hours before sunrise. After sunrise, O_3 mixing ratios increased to about 40 ppb at 12:00 h. With sunset around 18:00 h, O_3 mixing ratios started to decrease monotonically. Around 01:30 h the low night time values were reached again. ΔO_3 showed little variation over the day, remaining all the time at about $+3$ ppb. Only in the hours between 18:00 h and 01:30 h slightly higher ΔO_3 values were observed. The higher ΔO_3 values coincided with the period of decreasing O_3 mixing ratios. While the intensity of the turbulent transport decreased in the evening, the intensity of the O_3 sink at the ground must have remained high during the early evening hours, leading to an increase of ΔO_3. Because turbulent transport is stronger during day than during night, the relatively constant ΔO_3 points to a variable intensity of the O_3 sink at the surface. This is certainly caused by daytime stomatal (and non-stomatal) uptake of O_3. During nighttime (under low turbulence conditions) an additional sink, caused by the reaction of surface-near O_3 with NO (emitted from soil), should not be ruled out. The diel course of NO mixing ratio (Figure 6g) exhibited a peak between 06:30–08:30 h, reaching values of about 1.8 ppb. During the night hours, mixing ratios were typically below 0.5 ppb. However, no noticeable simultaneous peak was observed in ΔNO (Figure 6h). The reason for the NO peak was horizontal advection, which affected both measuring heights with the same magnitude. We will address the advection topic in more detail below. The diel course of ΔNO was highly variable during night (large inter-quartile range), while during day ΔNO was small, much less variable, but always negative (≈ -0.1 ppb). Microbial processes in the (top) soil are universal biogenic sources of NO, and at low background NO mixing ratios (a few to a few tens of ppb) nitric oxide is generally released from the soil (Conrad, 1996). This is in accordance with negative ΔNO during the entire LIBRETTO experiment. NO_2 mixing ratios (Figure 6i) exhibited a diel course similar to that of CO_2: highest values (4 ppb) during the second part of the night, a decrease to 1.8 ppb between 06:00 h and 12:00 h, quasi constant until 18:00 h and increasing again afterwards. The diel course of ΔNO_2, however, was different from that of ΔCO_2. With a short (and very small) exception around 19:30, the median ΔNO_2 was always positive throughout the day. However, during the afternoon and early evening hours, the observed inter-quartile range indicates small negative ΔNO_2 values. During the evening, median ΔNO_2 remained close to zero and eventually increased around midnight. Diel courses of air temperature T and ΔT (Figure 6k, 6l) followed the expected shape. T started to increase directly after sunrise and reached its highest values in the afternoon. Directly after sunset, T dropped rapidly, followed by a more gradual decrease during the rest of the night, cause by radiative energy loss. Positive ΔT at night indicated thermally stable conditions, while negative ΔT during the day confirmed thermally unstable conditions, enhancing turbulent transport.

Appendix C

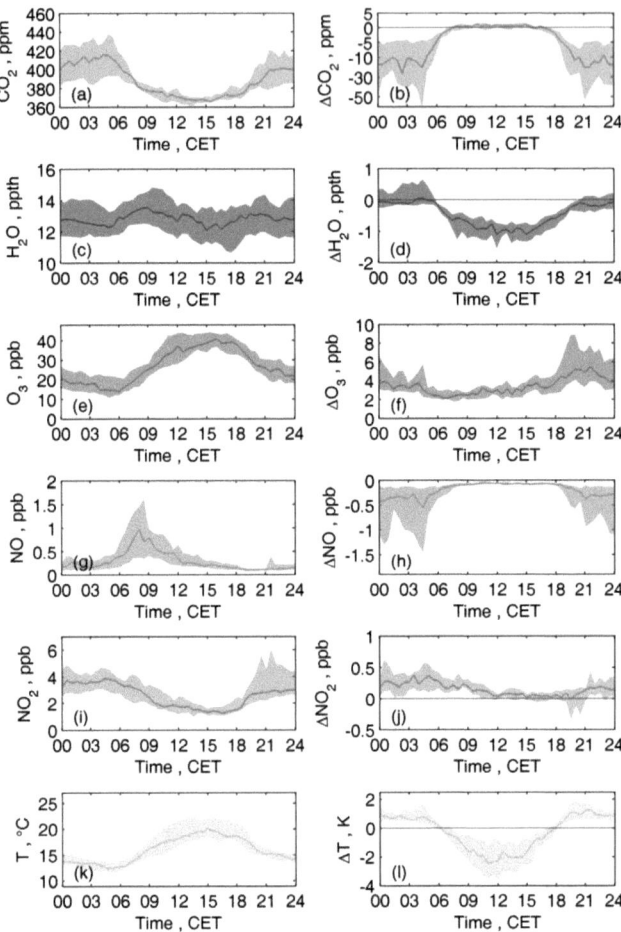

Figure 6: Median (straight lines) diel courses of CO_2 (olive), H_2O (blue), O_3 (red), NO (green), NO_2 (cyan) and T (orange), 11–30 August, 2006. Color shaded areas represent respective inter-quartile ranges. Left panels show the diel courses of the quantities measured at $z = 2$ m, right panels the difference between the measurements at $z = 2$ m and $z = 0.15$ m, respectively.

3.4.1 Advection

The median diel course of NO mixing ratios showed a peak between 06:30–08:30 h (see Section 3.4; Figure 6g). Because both measurement heights were affected by the same magnitude, advection of high NO mixing ratios, originating from traffic emissions, was suspected. Consequently, all days of the LIBRETTO campaign were analyzed for the occurrence of this NO peak. Out of 20 days, the peak was clearly identified on 7 days. On 6 days, definitely no peak was observed. Measurements during the remaining days were either influenced by local farming activities (hence, excluded from the analyses), or instationarities from the preceding night which obscured a clear identification. All peaks occurred between 07:00 h and 09:00 h.

Strong evidence for NO-advection from traffic emissions has been found: (a) no early-morning NO peak has been observed on the three Sundays covered by the LIBRETTO campaign, (b) at days with NO peak, corresponding air flow was from a different sector than that, from which the air flow arrived the site at days without the NO peak (see blue and red arrows in Figure 7), and (c) trucks operated by the beverage marked, located in Görsdorf (see Figure 7), left the market typically around 08:00 h, providing a strong local and temporally limited NO source. Moreover, the pale red sector (Figure 7), indicating airflows at days with early-morning NO peaks, contains a rural road which is relatively close to the Lindenberg experimental site (1.5 km and less). In contrast, the pale blue sector (indicating airflows at days lacking the NO peak) provides a much longer distance of the Lindenberg site to the next traffic related sources of NO.

Appendix C

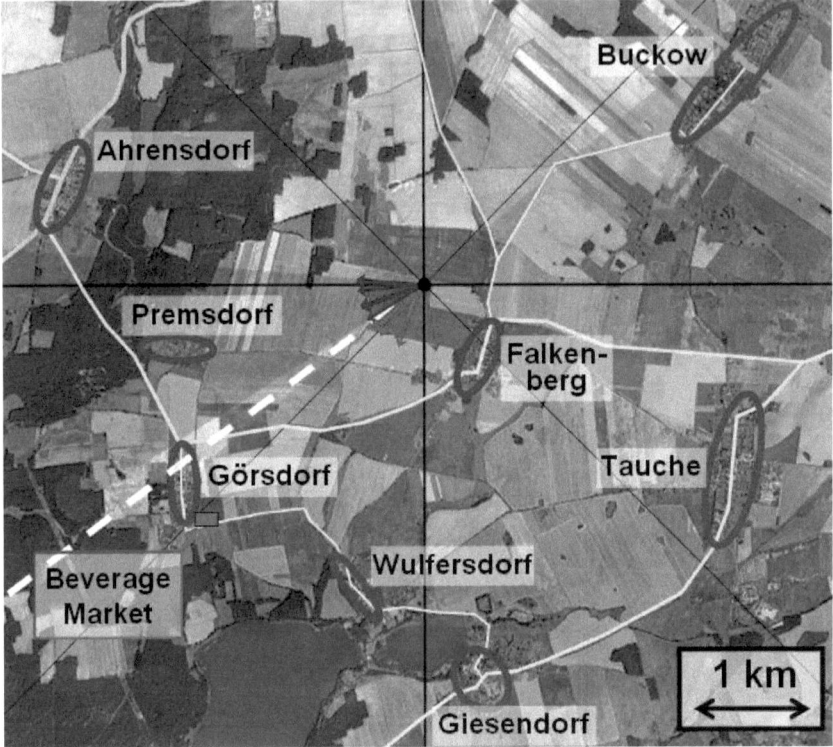

Figure 7: Aerial view of the surroundings of the Lindenberg field site. The sector highlighted in (pale) red comprises the directions of airflow (red arrows) for those days when the early-morning NO peak (see Figure 6g) was observed, while during days, lacking the NO peak, corresponding airflows (blue arrows) originated from the sector highlighted in (pale) blue.

3.5 Trace gases – fluxes

The trace gas fluxes, derived by the DMBR method, are shown in Figure 8. There, the convention is used, that downward (i.e. towards the surface) directed fluxes get a negative sign, upward fluxes are taken as positive. A clear diel cycle was visible in all trace gas fluxes except for NO. During the first part of the night, the observed CO_2-flux (Figure 8a) is characterized by peak respiration fluxes up to ≈ 5 $\mu mol\ m^{-2}\ s^{-1}$ (due to soil and plant respiration), while during the second part of the night, the CO_2-flux is gradually decreasing. Shortly after sunrise, CO_2-flux is changing sign, indicating the onset of photosynthesis. Maximum downward fluxes were reached at 12:00 h (-6 µmol m^{-2} s^{-1}). The median H_2O-flux (Figure 8b) was always positive, although very small towards the end of the night. Short periods of negative H_2O fluxes, i.e. dew fall, were possible then. With sunrise, the H_2O flux increased rapidly, reaching about $+1.8$ mmol m^{-2} s^{-1} around noon. This strong upward flux of H_2O was the result of plant transpiration and evaporation from soil surfaces. With sinking sun, the H_2O-flux decreased and eventually reached its low night time values of less than $+0.1$ mmol m^{-2} s^{-1} at 20:00 h. The flux of O_3 was found to be always directed towards the surface (Figure 8c). This was expected, as no surface source of O_3 is known, while there is stomatal up-take of O_3 and dry deposition onto surfaces (soil, plants) during daytime. As already mentioned, an additional O_3 sink at the surface (during conditions of low turbulence) is the reaction with NO, biogenically emitted from the soil. The median diel O_3-flux exhibited highest values in the early afternoon, peaking around -5.5 nmol m^{-2} s^{-1}. Elevated O_3-fluxes started with sunrise around 06:00 h and ended with sunset around 18:00 h. During the first part of the night, the O_3-flux remained around -2 nmol m^{-2} s^{-1} and decreased after midnight to about -0.8 nmol m^{-2} s^{-1}. In contrast to O_3, NO has a considerably strong source in the soil, hence positive (up-ward directed) NO-fluxes have been observed throughout the LIBRETTO campaign (Figure 8d). The NO-flux did not show a diel cycle, its median values ranged around $+0.1$ nmol m^{-2} s^{-1}. The median NO_2-flux remained negative during the entire day (Figure 8e), indicating net deposition of NO_2. A maximum deposition flux of about -0.2 nmol m^{-2} s^{-1} was observed in the morning hours, just after sunrise. Later, it leveled off and eventually reached zero around sunset. For the sake of completeness, the sensible heat flux H is also shown (Figure 8f). It exhibited very small and negative values during night (<-5 W m^{-2}). With sunrise, H increased rapidly, reaching median values of $+70$ Wm^{-2} around noon. In the afternoon, H decreased again and dropped below zero around 18:00 h, indicating the onset of surface cooling.

Appendix C

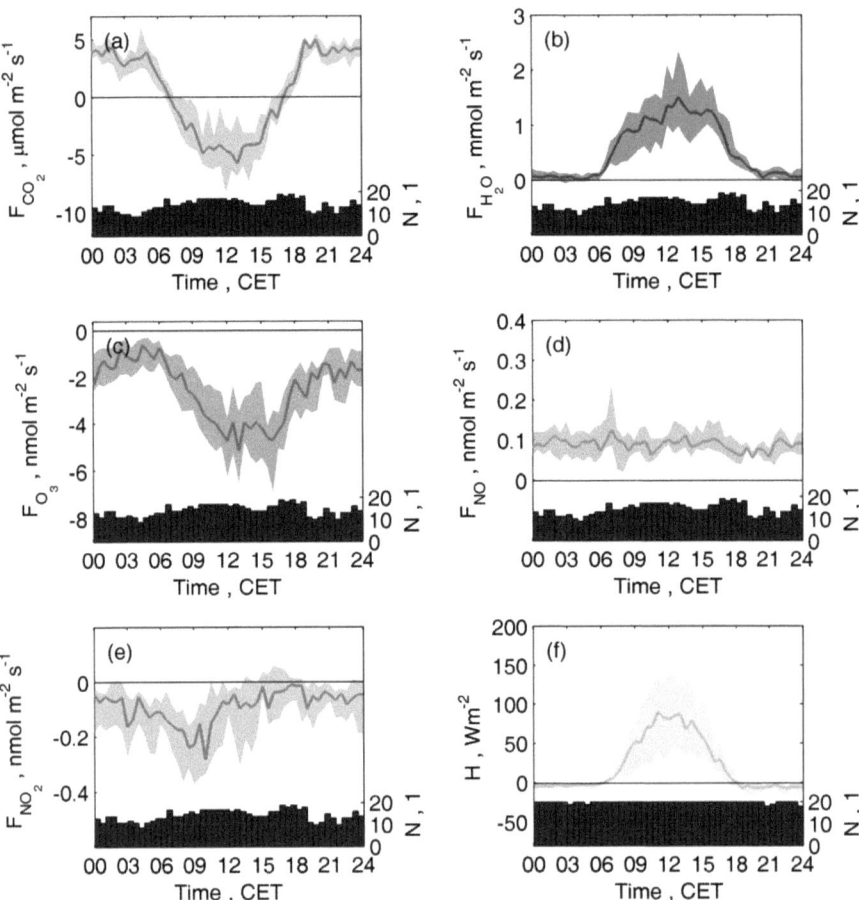

Figure 8: Median (straight lines) diel courses of CO2-flux (olive), H2O-flux (blue), O3-flux (red), NO-flux (green), NO2-flux (cyan), and sensible heat flux (orange), 11☐30 August, 2006. Color shaded areas represent respective inter-quartile ranges. Blue bars at the bottom of each graph indicate the number of data points available for calculation of medians and inter-quartile ranges.

3.5.1 Comparison of NO-fluxes: field vs. laboratory

Soil temperature and soil water content dependent net potential NO fluxes have been derived from laboratory incubation and fumigation experiments (see Section 2.4) for all soil samples taken the Lindenberg field site. Soil moisture and soil surface temperature data obtained from field measurements during the LIBRETTO campaign were used to generate a "laboratory derived NO-flux" from corresponding net potential NO fluxes. The obtained time series of "laboratory derived NO-fluxes" was then converted into a median diel cycle, analogously to the median diel cycle of NO-fluxes derived by the DMBR method. Resulting diel courses are shown in Figure 9.

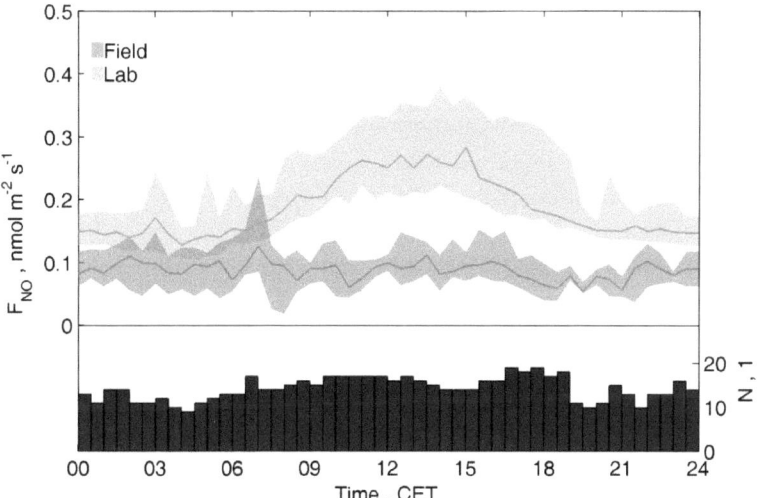

Figure 9: Median diel courses of NO-flux from field measurements (green) and from up-scaled (laboratory derived) net potential NO-fluxes (grey); up-scaling was achieved with field data of soil moisture and soil surface temperature. Straight lines represent the medians of NO-fluxes, while color shaded areas represent their corresponding inter-quartile ranges; blue bars at the bottom indicate the number of data points available for calculation of medians and inter-quartile ranges.

Appendix C

The laboratory derived NO-fluxes exceeded the field NO-fluxes by a factor of 1.5 to 2.5. While a considerable diel amplitude was observed for the laboratory derived NO-flux, the field NO-flux remained almost constant around + 0.1 nmol m^{-2} s^{-1}. Acknowledging the very different nature of both methods to derive NO-fluxes, the agreement between the two data sets is considered to be quite good. Nevertheless, we like to point to one aspect which might have caused the observed deviation between the two methods. The laboratory derived NO-flux is *sensu stricto* valid only at the immediate surface of the soil ($z = 0$ m; c.f. (Galbally and Johansson, 1989), while the field NO-flux is attributed to $z_m = 0.55$ m above ground $z_m = (z_1 \cdot z_2)^{0.5}$, z_1=0.15 m, z_2=2.0 m, see Eq. (1)). Because turbulence becomes very weak close to the ground, the residence time of NO in the layer between soil surface ($z = 0$ m) and $z_m = 0.55$ m might become long (compared to the characteristic time of the NO–O$_3$ reaction (R1)). Therefore, a considerable vertical divergence of the NO-flux can occur, which would reduce the NO-flux with increasing distance from the soil surface to $z_m = 0.55$ m. In this context, it should be mentioned, that the assumption to treat NO as a quasi passive tracer (see Section 3.2) is justified (by Damköhler number arguments) only for the considered layer between the two measurement heights (z_1=0.15 m, z_2=2.0 m).

3.5.2 Comparison of methods: DMBR vs. NBLB (nocturnal boundary layer budget)

The coexistence of surface layer flux measurements of CO_2, O_3 (DMBR method), as well as H (EC method), and highly resolved vertical profiles of CO_2, O_3 and T up to 100 m provided the option to compare the surface layer fluxes with fluxes derived by the NBLB method.

One night during the LIBRETTO campaign provided sufficient atmospheric stability and stationarity for this comparison. Time-height cross sections of potential temperature (θ), CO_2 and O_3 are shown in Figure 10. After the transition of the sensible heat flux to negative values (left vertical dashed line, Figure 10), which indicates radiative cooling, the entire air column started to cool down from about 23 °C to approx. 16 °C within 3 hours, while the strongest cooling was at the beginning of this period. Later during the night, the air was cooled by the surface, which in turn was cooled by outgoing long-wave radiation. Shortly before 06:00 h in the morning, the sensible heat flux changed to positive values, which indicates the start of surface heating. This became immediately visible in the profiles of θ, showing unstable thermal stratification near the ground. The situation for CO_2 was somewhat different. After the atmosphere became stable, CO_2 enrichment started from close to the ground (Figure 10b). The CO_2 mixing ratio close to the ground increased with time, while the layer with increased CO_2 mixing ratios (in comparison to the afternoon values), was deepening to an extent, which eventually exceeded the range covered by the mast data. Highest mixing ratios occurred on the second half of the night, exceeding 410 ppm (at $z = 2$ m). When sensible heat flux was changing back to positive values shortly before 06:00 h in the morning, CO_2 mixing ratio began to drop back to lower values simultaneously in the entire observable column. This was the result of the zero crossing of the CO_2 flux (from upward to downward directions, see Figure 8a) and the simultaneous growing of the mixing layer. The time-height cross section of O_3 (Figure 10c) shows small positive vertical gradients of O_3 mixing ratio during day with only small temporal changes of the mixing ratios. After 19:30 h, O_3 mixing ratios decreased fast at all heights, while the vertical gradients increased. This was primarily the effect of reduced turbulence, with only slow vertical transport velocities. It led to a loss of O_3 in the entire observable column, however, being strongest close to the ground. Low O_3 mixing ratios prevailed in the observable column until approx. two hours after sunrise. During these two hours, only O_3 depleted air from the former NBL was entrained into the newly developing boundary layer. At 08:00 h, O_3 mixing ratios started to increase, an indication that now air from the residual layer (not affected by nocturnal O_3 depletion) was entrained.

Appendix C

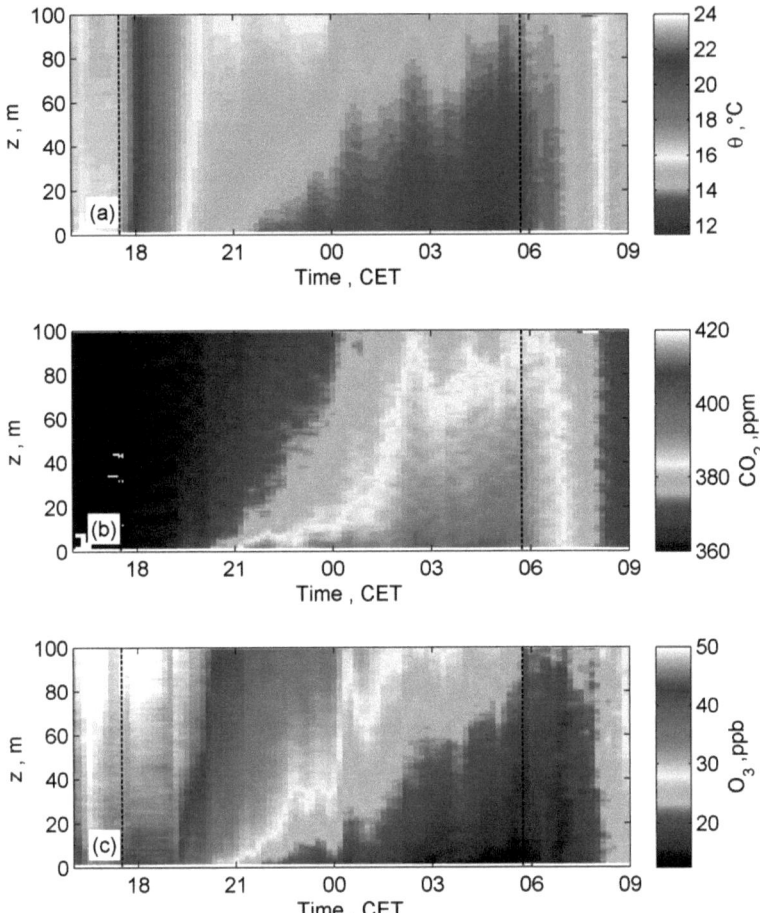

Figure 10: Time-height cross section of (a) potential temperature, (b) CO_2 mixing ratio and (c) O_3 mixing ratio for 20 August 2006 16:00 h to 21 August 2006 09:00 h. The dashed vertical lines denote the zero crossing of the sensible heat flux, measured at $z = 2.4$ m, being negative during night.

A direct comparison of the surface layer fluxes (DMBR and EC method) with the fluxes derived with the NBLB method revealed further nocturnal dynamics (see Figure 11). The time series of the fluxes (Figures 11 a, c, d) exhibited a close correlation between the DMBR and the NBLB method. The magnitude of the fluxes as well as the time of their zero crossing agreed well. Only H showed major differences before 20:00 h. Nevertheless, fluxes derived with the NBLB method had a larger temporal variability. A different way of comparing the temporal evolution of the fluxes is a plot of the cumulative flux of heat or matter into or out of the NBL. For the cumulative heat flux (Figure 11b), two nocturnal episodes can be discerned: an initial stabilization phase, and a subsequent cooling phase for the remainder of the night. During the first part until 20:30 h, H measured by the EC method yielded only $-$ 12.8 J m^{-2} h^{-1} while H from the NBLB yielded $-$ 104.6 J m^{-2} h^{-1}. Such a discrepancy was already suggested by Figure 10a, where an initial rapid cooling throughout the entire observable air column was visible. It could be explained by the adjustment of the air temperature to the altered radiation regime after sunset. Consequently, because this temperature decrease occurred *"in situ"*, the H from EC measurements did not reflect the magnitude of heat change. In the second part of the night, from 20:30 h until sunrise, both methods yielded similar values of H ($-$ 6.1 J m^{-2} h^{-1} by EC and $-$ 7.7 J m^{-2} h^{-1} by NBLB methods, respectively). The separation of nocturnal dynamics into a first part before 20:30 h and a second part after 20:30 h (until sunrise) was also observable in the trace gas flux data. The cumulative flux of CO_2 was similar from both methods in the second part (DMBR: + 13.5 mmol m^{-2} h^{-1}, NBLB: + 11.9 mmol m^{-2} h^{-1}). It indicated that the temporal change of CO_2 mixing ratios within the NBL was only controlled by the surface respiration. All CO_2 accumulating in the NBL had thus to pass the DMBR measuring level. In the first part of the night, however, a slight deviation between the two methods was observed. The NBLB method yielded slightly higher CO_2 fluxes (+ 25.1 mmol m^{-2} h^{-1}) than the DMBR method (+ 12.9 mmol m^{-2} h^{-1}). A very different situation was found for the O_3-flux. In the second part of the night, the cumulative O_3-fluxes showed revealed great differences between the NBLB ($-$ 8.3 µmol m^{-2} h^{-1}) and the DMBR derived O_3-fluxes ($-$ 2.5 µmol m^{-2} h^{-1}). Here, the striking difference of O_3 compared to CO_2 and sensible heat has to be considered, namely that the fact, that O_3 is simply a reactive trace gas. Because vertical exchange was suppressed during night, even relatively slow chemical reactions (R1) led to a depletion of O_3. This does

Appendix C

not only occur at the surface, but happened also in the entire observable column, i.e. within the NBL. Photolytic reformation of O_3 from NO_2 (R2) does not occur during night. Furthermore, additional O_3 might have been destroyed by reaction with NO_2, forming the NO_3 radical (Seinfeld and Pandis, 1998). However, the DMBR method captured only the O_3-flux directed towards the surface. In contrast, the NBLB was affected additionally by *in-situ* destruction of O_3 within the NBL, leading to a temporal change and thus an apparent flux. While the NBLB did obviously not provide correct flux data in case of reactive trace gases, a comparison between the two methods can give an estimate about the magnitude of in-situ chemical O_3 loss within the NBL.

Finally it has to be mentioned in this context, that the fluxes derived by the NBLB did not only represent the immediate field site, as the surface layer fluxes (by DMBR) do. Due to the large fetch, different types of land use upwind of the mast might have affected the profiles and thus the actual temporal change of a variety of scalars (Beyrich et al., 2002).

Appendix C

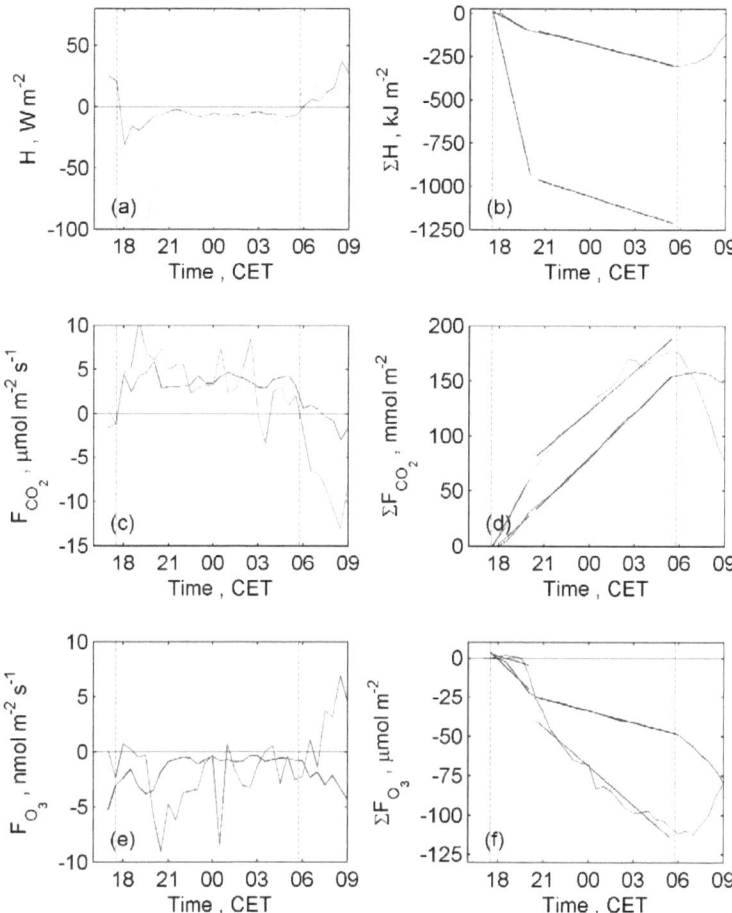

Figure 11: Comparison of fluxes of sensible heat (H), CO2 (FCO2) and O3 (FO3) measured with the DMBR or EC method (dark colors) and derived by the NBLB (light colors). Panels (a), (c) and (e) show the 30 minute values of the respective fluxes, while panels (b), (d) and (f) show the cumulative flux since onset of surface cooling (indicated by the left dashed line) The right dashed line marks the onset of surface heating after sunrise.

4 Conclusions

We have demonstrated, as long as the terrain is homogeneous with respect to roughness and soil/vegetation properties, that the MBR method can be used in its distributed version (DMBR), where sensible heat flux and corresponding temperature gradient are measured a few tens of meters apart from trace gas gradient measurements. This provides the option of using one EC station for several, distributed trace gas flux measurements at one site.

If the DMBR method is applicable, it provides a quick and reliable way to compute vertical fluxes of trace gases in the surface layer, as it was demonstrated for the passive trace gases CO_2 and H_2O, as well as for the reactive trace gases O_3, NO and NO_2. However, it has to be noted, that such fluxes might not be necessarily equal to the fluxes at the immediate soil surface, nor equal to the fluxes as they are derived from the integration of (reactive) trace gas profiles up to their equilibrium level. Under stable conditions, with strong surface inversions, trace gases are trapped already close to the ground, leading there to very high mixing ratios. The observed surface layer fluxes represent (a) the mean flux for the layer limited by the measurement heights to establish the mixing ratio differences, and (b) only the outflow from (or inflow into) this pool of trapped trace gas. But still, they represent the lower flux boundary condition for (reactive) trace gas profiles above.

The comparison of the fluxes derived with the DMBR method and from integration of the respective trace gas profiles up to 100 m (NBL budget) revealed very clearly the different spatial domains of both approaches. While the DMBR method provided the local vertical turbulent flux in the surface layer (precisely: between the two levels of the mixing ratio measurements), the profile integration covered a large footprint, i.e. a large horizontal area from which the fluxes originated. The principal differences between the fluxes of the conservative quantities CO_2 and sensible heat points out, that transport dynamics of temperature and trace gases may differ significantly. Furthermore, with increasing height increasing lateral fetch heterogeneity of the surface is included into the profiles. This leads to variations, not representative for the surface directly underlying the profile.

Acknowledgements

The authors are very grateful to the German Meteorological Service, especially to the staff of the Richard-Aßmann-Observatory in Lindenberg, for allowing the use of their Boundary Layer Field Site Falkenberg and for providing all the supporting reference data. Special thanks belong to the members of the LIBRETTO crew (Monika Scheibe, Korbinian Hens and Michael Kröger), whose dedication made this campaign possible. The project was funded by the Max Planck Society (Germany).

Appendix C

References

Akima, H.: A new method of interpolation and smooth curve fitting based on local procedures, Journal of the Association for Computing Machinery, 17, 589-602, 1970.

Arya, S. P. S.: Introduction to Micrometeorology, Intern. Geophys. Ser., Academic Press, San Diego, CA., 420 pp., 2001.

Atkinson, R., Baulch, D. L., Cox, R. A., Crowley, J. N., Hampson, R. F., Hynes, R. G., Jenkin, M. E., Rossi, M. J., and Troe, J.: Evaluated kinetic and photochemical data for atmospheric chemistry: Volume I - gas phase reactions of Ox, HOx, NOx and SOx species, Atmos. Chem. Phys., 4, 1461-1738, 2004.

Aubinet, M., Grelle, A., Ibrom, A., Rannik, Ü., Moncrieff, J., Foken, T., Kowalski, A. S., Martin, P. H., Berbigier, P., Bernhofer, C., Clement, R., Elbers, J., Granier, A., Grunwald, T., Morgenstern, K., Pilegaard, K., Rebmann, C., Snijders, W., Valentini, R., and Vesala, T.: Estimates of the annual net carbon and water exchange of forests: The EUROFLUX methodology, in: Advances in Ecological Research, Vol 30, Advances in Ecological Research, ACADEMIC PRESS INC, San Diego, 113-175, 2000.

Baldocchi, D., Falge, E., Gu, L. H., Olson, R., Hollinger, D., Running, S., Anthoni, P., Bernhofer, C., Davis, K., Evans, R., Fuentes, J., Goldstein, A., Katul, G., Law, B., Lee, X. H., Malhi, Y., Meyers, T., Munger, W., Oechel, W., U, K. T. P., Pilegaard, K., Schmid, H. P., Valentini, R., Verma, S., Vesala, T., Wilson, K., and Wofsy, S.: FLUXNET: A new tool to study the temporal and spatial variability of ecosystem-scale carbon dioxide, water vapor, and energy flux densities, Bulletin of the American Meteorological Society, 82, 2415-2434, 2001.

Baldocchi, D. D., Hicks, B. B., and Meyers, T. P.: Measuring Biosphere-Atmosphere Exchanges of Biologically Related Gases with Micrometeorological Methods, Ecology, 69, 1331-1340, 1988.

Bargsten, A., Falge, E., Huwe, B., and Meixner, F. X.: Laboratory measurements of nitric oxide release from forest soil with a thick organic layer under different understory types, Biogeosciences Discussions, 7, 203-250, 2010.

Beyrich, F., Richter, S. H., Weisensee, U., Kohsiek, W., Lohse, H., de Bruin, H. A. R., Foken, T., Gockede, M., Berger, F., Vogt, R., and Batchvarova, E.: Experimental determination of turbulent fluxes over the heterogeneous LITFASS area: Selected results from the LITFASS-98 experiment, Theoretical and Applied Climatology, 73, 19-34, 2002.

Beyrich, F., and Adam, W. K.: Site and Data Report for the Lindenberg Reference Site in CEOP – Phase I, Deutscher Wetterdienst, Offenbach230, 55, 2007.

Bowen, I. S.: The Ratio of Heat Losses by Conduction and by Evaporation from any Water Surface, Physical Review, 27, 779-787, 1926.

Businger, J. A., Wyngaard, J. C., Izumi, Y., and Bradley, E. F.: Flux-profile relationships in the atmospheric surface layer, Journal of Atmospheric Sciences, 28, 181-189, 1971.

Businger, J. A.: Evaluation of the Accuracy with which Dry Deposition can be Measured with Current Micrometeorological Techniques, Journal of Climate and Applied Meteorology, 25, 1100-1124, 1986.

Conrad, R.: Soil microorganisms as controllers of atmospheric trace gases (H2, CO, CH4, OCS, N2O, and NO), Microbiological Reviews, 60, 609-640, 1996.

Damköhler, G.: Der Einfluss der Turbulenz auf die Flammengeschwindigkeit in Gasgemischen, Zeitschrift für Elektrochemie und Angewandte Physikalische Chemie, 46, 601-652, 1940.

Denmead, O. T., Raupach, M. R., Dunin, F. X., Cleugh, H. A., and Leuning, R.: Boundary layer budgets for regional estimates of scalar fluxes, Global Change Biology, 2, 255-264, 1996.

Eugster, W., and Siegrist, F.: The influence of nocturnal CO_2 advection on CO_2 flux measurements, Basic and Applied Ecology, 1, 177-188, 2000.

Feig, G. T., Mamtimin, B., and Meixner, F. X.: Soil biogenic emissions of nitric oxide from a semi-arid savanna in South Africa, Biogeosciences, 5, 1723–1738, 2008.

Foken, T., Dlugi, R., and Kramm, G.: On the determination of dry deposition and emission of gaseous compounds at the biosphere-atmosphere interface, Meteorologische Zeitschrift, 4, 91-118, 1995.

Foken, T.: Micrometeorology, Springer, Heidelberg, New York, 310 pp., 2008.

Galbally, I. E., and Johansson, C.: A model relating laboratory measurements of rates of nitric oxides production and field measurements of nitric oxide emission from soils Journal of Geophysical Research, 94, 6473-6480, 1989.

Högström, U.: Non-dimensional wind and temperature profiles in the atmospheric surface layer: A re-evaluation., Boundary-Layer Meteorology, 42, 55-78, 1988.

Keronen, P., Reissell, A., Rannik, Ü., Pohja, T., Siivola, E., Hiltunen, V., Hari, P., Kulmala, M., and Vesala, T.: Ozone flux measurements over a scots pine forest using eddy covariance method: performance evaluation and comparison with flux-profile method, Boreal Environment Research, 8, 425-443, 2003.

Lenschow, D. H.: Reactive trace species in the boundary layer from a micrometeorological perspective, Journal of the Meteorological Socienty of Japan, 60, 472-480, 1982.

Levy, P. E., Grelle, A., Lindroth, A., Mölder, M., Jarvis, P. G., Kruijt, B., and Moncrieff, J. B.: Regional-scale CO_2 fluxes over central Sweden by a boundary layer budget method, Agricultural and Forest Meteorology, 98-99, 169-180, 1999.

Lewis, J. M.: The Story behind the Bowen Ratio, Bulletin of the American Meteorological Society, 76, 2433-2443, 1995.

Liu, H., and Foken, T.: A modified Bowen ratio method to determine sensible and latent heat fluxes, Meteorologische Zeitschrift, 10, 71-80, 2001.

Mathieu, N., Strachan, I. B., Leclerc, M. Y., Karipot, A., and Pattey, E.: Role of low-level jets and boundary-layer properties on the NBL budget technique, Agricultural and Forest Meteorology, 135, 35-43, 2005.

Mayer, J.-C., Hens, K., Rummel, U., Meixner, F. X., and Foken, T.: Moving measurement platforms - specific challenges and corrections, Meteorologische Zeitschrift, 18, 477-488, 2009.

Müller, H., Kramm, G., Meixner, F., Dollard, G. J., Fowler, D., and Possanzini, M.: Determination of HNO_3 Dry Deposition by Modified Bowen-Ratio and Aerodynamic Profile Techniques, Tellus Series B, 45, 346-367, 1993.

Ormeci, B., S.L., S., and Peirde, J. J.: Laboratory study of NO flux from agricultural soil: Effects of soil moisture, pH, and temperature, Journal of Geophysical Research, 104, 1621-1629, 1999.

Pattey, E., Strachan, I. B., Desjardins, R. L., and Massheder, J.: Measuring nighttime CO2 flux over terrestrial ecosystems using eddy covariance and nocturnal boundary layer methods, Agricultural and Forest Meteorology, 113, 145-158, 2002.

Paulson, C. A.: The Mathematical Representation of Wind Speed and Temperature Profiles in the Unstable Atmospheric Surface Layer, Journal of Applied Meteorology, 9, 857-861, 1970.

Rohrer, F., Brünin, D., Grobler, E. S., Weber, M., Ehhalt, D. H., Neubert, R., Schüßler, W., and Levin, I.: Mixing Ratios and Photostationary State of NO and NO2 Observed During the POPCORN Field Campaign at a Rural Site in Germany, Journal of Atmospheric Chemistry, 31, 119-137, 1998.

Rummel, U., Ammann, C., Gut, A., Meixner, F. X., and Andreae, M. O.: Eddy covariance measurements of nitric oxide flux within an Amazonian rain forest, Journal of Geophysical Research, 107, 8050, doi:10.1029/2001JD000520, 2002.

Seinfeld, J. H., and Pandis, S. N.: Atmospheric Chemistry and Physics: From Air Pollution to Climate Change, Wiley, New York, 1998.

Suni, T., Rinne, J., Reissell, A., Altimir, N., Keronen, P., Rannik, Ü., Dal Maso, M., Kulmala, M., and Vesala, T.: Long-term measurements of surface fluxes above a Scots pine forest in Hyytiala, southern Finland, 1996-2001, Boreal Environment Research, 8, 287-301, 2003.

Trebs, I., Bohn, B., Ammann, C., Rummel, U., Blumthaler, M., Koenigstedt, R., Meixner, F. X., Fan, S., and Andreae, M. O.: Relationship between the NO2 photolysis frequency and the solar global irradiance, Atmos. Meas. Tech. Discuss., 2, 1537-1573, 2009.

van Dijk, S. M., and Meixner, F. X.: Production and Consumption of NO in Forest and Pasture Soils from the Amazon Basin, Water, Air, & Soil Pollution, 1, 119-130, 2001.

Vickers, D., and Mahrt, L.: Quality Control and Flux Sampling Problems for Tower and Aircraft Data, Journal of Atmospheric and Oceanic Technology, 14, 512-526, 1997.

Vilà-Guerau de Arellano, J., and Duynkerke, P. G.: Influence of chemistry on the flux-gradient relationships for the $NO-O_3-NO_2$ system, Boundary-Layer Meteorology, 61, 375-387, 1992.

Wesely, M. L., Eastman, J. A., Stedman, D. H., and Yalvac, E. D.: An eddy-correlation measurement of NO_2 flux to vegetation and comparison to O_3 flux, Atmospheric Environment, 16, 815-820, 1982.

Wyngaard, J. C.: Boundary-Layer Modelling, in: Atmospheric Turbulence and Air Pollution Modelling, edited by: Nieuwstadt, F. T. M., and van Dop, H., Reidel, Dordrecht, The Netherlands, 69-106, 1982.

Yu, J., Meixner, F. X., Sun, W., Liang, Z., Chen, Y., Mamtimin, B., Wang, G., and Sun, Z.: Biogenic nitric oxide emission from saline sodic soils in a semiarid region, northeastern China: A laboratory study, J. Geophys. Res., 113, G04005, doi:04010.01029/02007JG000576, 2008.

Appendix D

Low nitrous oxide emissions in an unmanaged old growth beech forest

A. Bargsten[1], M. Timme[2,3], S. Glatzel[4] and H. F. Jungkunst[5]

[1] Biogeochemistry Department, Max Planck Institute of Chemistry, 55020 Mainz, Germany

[2] Network Dynamics Group, Max Planck Institute for Dynamics & Self-Organization, Göttingen, Germany

[3] Faculty of Physics, University of Göttingen, Germany

[4] Landscape Ecology and Site Evaluation, Faculty of Agricultural and Environmental Sciences, University of Rostock, 18059 Rostock, Germany

[5] Landscape Ecology, Institute of Geography, University of Göttingen, Germany

In preparation for the European Journal of Soil Science

Appendix D

Abstract

Nitrous oxide (N_2O) is a high-impact greenhouse gas, but its long-term natural background emission level is not well understood. Here we measured soil N_2O emissions in an unmanaged, old growth beech forest in the Hainich National Park, Germany, at 15 plots over a one-year period. The average annual field N_2O flux rate was 0.46±0.32 kg ha^{-1} yr^{-1} (in terms of mass of N). The N_2O emissions show a background emission pattern with two event based N_2O peaks. A correlation analysis shows that the distance between plots (up to 380 m) is secondary for their flux correlations. Annual N_2O fluxes obtained from a standard model (Forest-DNDC) given soil parameters as well as daily temperature and precipitation substantially overestimate the actual field N_2O fluxes and also do not describe their actual temporal and spatial variabilities. Temporal variability was described well by the model only at plots with higher soil organic carbon and the modelled N_2O fluxes increased during freezing only were soil organic carbon was larger than 0.06 kg C/kg. The results indicate that the natural background of nitrous oxide emissions may be lower than previously thought and also lower than assumed in standard modeling. This suggests a higher anthropogenic contribution to N_2O emissions. Moreover, standard flux models need to be revised to study natural emissions, in particular with respect to mechanisms underlying N_2O emissions.

1 Introduction

The atmospheric abundance of the greenhouse gas nitrous oxide (N_2O) has been rising since industrialization and intensification of agriculture and anthropogenic emissions need to be reduced again to counteract global warming (Denman *et al.*, 2007). Consequently, anthropogenic emissions need to be separated from natural background emissions for inventories and for the scientific understanding. This is not a trivial task and to date a background emission of 1.0 kg ha^{-1} yr^{-1} (in terms of mass of N) is commonly used as a standard (Bouwman, 1996). This estimate derives from measurements of unfertilized ecosystems. However, most so-called unfertilized sites are not sites that were never fertilized. Furthermore, the time since the last fertilization varies considerably and frequently these sites were solely not fertilized during the measurement years but the years just before measurements started. Mostly, N-fertilization is used as it is considered as one of the main drivers for N_2O emissions (Skiba and Smith, 2000). This is problematic as anthropogenic influence is not restricted to N-fertilization (Ruser et al., 1998; Teepe et al., 2004; Flessa et al., 2002b) and plowing (Mosier et al., 1996), but it is a very pragmatic and reasonable approach. As it is still unclear after which period of withholding fertilization emission rates have returned back to their natural background level, it is possible that these "recently unfertilized" systems still exhibit elevated N_2O caused by fertilization before the withholding period. This and other natural and anthropogenic influences, such as atmospheric N-depositions and soil compaction, may lead to biased estimates of the natural background emission rates.

Actual natural background fluxes are those of the undisturbed ecosystem that are unmanaged and exhibit vegetation close to the potential natural one. However, such ecosystems have become very rare across the globe (Groombridge and Jenkins, 2000). For the greater part of Central Europe, the potential natural vegetation is beech (*Fagus sylvatica*) forest (Ellenberg, 1996). Hence natural background emission rates representative for Central Europe should be recorded in pristine beech forests preferably at contrasting sites, e.g. at different soils and under different climate. However, there are hardly any unmanaged beech forests and there are certainly no pristine beech forests left in Central Europe. Nevertheless, in unmanaged old forest ecosystems anthropogenic influences may be assumed to be low because the N-status should not elevated besides the unavoidable atmospheric deposition.

Ecosystems in the core zone of the Hainich National Park (NP) is are among the closest to natural available in all of Central Europe and thus provide the opportunity to determine the closest estimate to the potential natural background emission in Central Europe.

Furthermore, if N_2O release can be inferred from an "uncalibrated" biogeochemical model, the model can be used to determine a proxy for close to natural N_2O fluxes for similar ecoregions in Central Europe. By using Forest-DNDC (http://www.dndc.sr.unh.edu/, January 2010) with the default parameter settings, this assessment would also be applicable to such regions. The outcome of measurement and modelling approach of a long-term unmanaged beech forest will add to the understanding of N_2O fluxes from natural ecosystems.

A well known phenomenon associated with gas flux measurement from soils is the high spatial and temporal variability (Folorunso and Rolston, 1984). An issue that needs to be considered in this respect is the question of spatial correlation (Jordan et al., 2009). What is the optimal distance between flux measurement chambers to avoid strong correlation (e.g. pseudo-replication)?

The objectives of this study were: (a) to determine the N_2O flux rate of an unmanaged beech forest site in the Hainich National Park for one year measured at 15 plots and to determine their spatial correlations; (b) to determine the standard soil parameters at the 15 plots in order to (c) make 15 uncalibrated runs with the model Forest-DNDC (Li et al., 2000; Butterbach-Bahl et al., 2001) and (d) to compare the 15 modelled N_2O flux rates with the 15 measured flux rates. (e) Thereby the relevance of spatial variability and the performance of Forest-DNDC for the assessment of N_2O fluxes in unmanaged temperate forests will be elucidated.

2 Materials and Methods

2.1 Study site

Our research site is located within the Hainich National Park (51°04'46''N, 10°27'08''E, 440 m a.s.l.) in Thuringia, Germany. The Hainich National Park was established in 1997 to protect one of the largest beech forests in Central Europe. Due to a unique history as a military base for more than 60 years prior to 1997, a large part of the forest has been taken out of management and developed with little disturbance. In the centuries before, the forest at the Hainich research site was used by the local village population as a coppice with standard systems and therefore has not been exposed to clearcut (Gleixner et al., 2009). As a consequence, the trees cover a wide range of age classes with a maximum up to 250 years. The forest is dominated by beech (Fagus sylvatica, 65%). The above-ground stem carbon pool is about 130 t C ha^{-1} (Gleixner et al., 2009). Maximum tree height varies between 30 and 35 m with a maximum leaf area index (LAI) of 5.0 m^2 m^{-2}. The long-term mean annual air temperature is 7.5-8°C and the mean annual precipitation is 750-800 mm.

2.2 Field measurements and N_2O flux analysis

N_2O flux measurements were carried out using a closed chamber technique. 15 cylindrical polyvinyl chloride (PVC) frames (30 cm in diameter and 15 cm tall) were installed at the research site. The frames were set up in the topsoil two weeks before starting the gas sampling. The location of each frame was selected stratified randomly. The largest distance between the frames was 380 m.

N_2O fluxes was measured 34-times (resulting in 34x15=510 N_2O flux measurements) within the period from November 2005 to November 2006 by placing a PVC lid (30 cm in diameter and 30 cm tall) at each frame and taking five gas samples from the chamber headspace using gas tight syringes (60 ml) after 0, 10 and 20 minutes of closure. N_2O concentration were analyzed in the laboratory using an automated gas chromatograph (GC) system (GC-14B, Shimadzu, Germany) equipped with flame ionization and an electron capture detector. A detail description of the GC system is given by Loftfield et al. (1997). For calibration three certified standards were used (303 ppb, 1000 ppb, 1998 ppb). N_2O fluxes were calculated using the linear regression of gas concentration versus time for each chamber. N_2O fluxes were rejected if the regression coefficient (r^2) fell below 0.7, because this indicates for e.g. that the chamber was not properly sealed, or that air samples were somehow contaminated. Mean N_2O fluxes were calculated using the N_2O fluxes with $r^2 > 0.7$. All N_2O fluxes were given in terms of mass of N.

2.3 Forest-DNDC

Forest-DNDC simulates C and N dynamics in soil as well as trace gas emissions (like N_2O, CH_4, N_2, NO and NH_3) from wetland and upland forested ecosystems. Biological processes are driven by climate, soil biogeochemistry, vegetation, and anthropogenic activity. The general structure of the model was adopted from PnET-N-DNDC, which simulates C and N dynamics of upland soils (Stange, 2001; Li et al., 2000). The model can be run in the upland mode, which is the same as PnET-N-DNDC, or in the wetland mode. The model has been successfully tested and applied for both, the upland and wetland mode (Stange, 2001; Stange et al., 2000). The PnET-N-DNDC model was constructed by integrating a series of new developments with three existing models, namely, the Photosynthesis-Evapotranspiration (PnET) model, the Denitrification-Decomposition (DNDC) model, and the nitrification model (Li et al., 2000). A more detailed description of the Forest-DNDC model is given in Stange et al. (2000) and Li et al. (2000) (see also Fig. 1).

In our study, we ran the Forest-DNDC model (http://www.dndc.sr.unh.edu/, download: January 2010) for all 15 plots in the upland mode to simulate the N_2O fluxes for the Hainich research site. Meteorological input data, required as model drivers, were minimum and

Appendix D

maximum daily air temperature as well as the sum of daily precipitation. Model input parameter were forest type, forest age, above ground biomass and soil parameters (mineral soil, organic layer, soil pH, soil organic carbon, bulk density, clay content) (see Tab. 1). Otherwise we use the default parameter settings.

Table 1: Overview of the model input parameter

	Model input parameter
forest type	beech
forest age	120 years
above ground biomass	130 t C ha^{-1}
mineral soil	clay loam
organic layer	mull
soil pH	individual for each plot (see Tab. 3)
soil organic carbon	individual for each plot (see Tab. 3)
bulk density	individual for each plot (see Tab. 3)
clay content	individual for each plot (see Tab. 3)

The modelled N_2O fluxes will be present on two ways:

- in daily resolution for the whole measuring period (model (daily)) and
- modelled N_2O fluxes for the 34 measuring days (model).

Appendix D

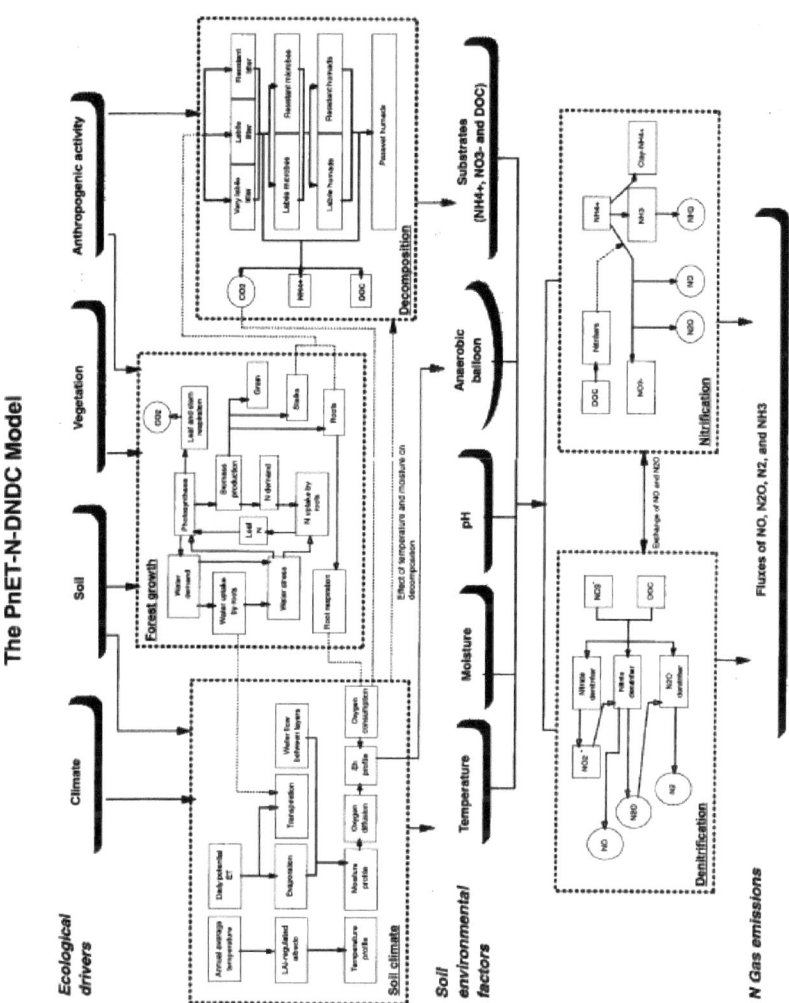

Figure 1: The PnET-N-DNDC model consists of five submodels for predicting soil climate, forest growth, decomposition, nitrification, and denitrification. The first three submodels from a component to calculate soil climate/substrate profiles driven by the ecological drivers, and the last two submodels from another component to predict nitrification, denitrification, and chemodenitrification reates based on soil environmental conditions (Li et al., 2000).

Appendix D

2.4 Climate and soil parameter

Meteorological data were observed at a station located outside the forest installed by the Max Planck Institute for Biogeochemistry.

Additionally, we determined soil temperature and soil moisture at each plot at the time of gas sampling. Soil temperature was measured in 12 cm depth using a mobile temperature sensor (Testo 110, Testo, Germany). Soil moisture was determined gravimetrically from a sample of the top 10 cm of soil by relating the mass difference between the fresh soil sample and the afterwards oven dried soil sample (105°C for 24 h) to the mass of the dry soil sample. To compare the gravimetric soil water content with the soil moisture displayed by the Forest-DNDC model (see next subsection) we calculated the water filled pore space (WFPS) according to Parton et al. (2001) by

$$WFPS = \theta * \left(\frac{BD}{WD}\right) * \left(\frac{PD}{PD-BD}\right) \qquad (1)$$

where θ is the gravimetric soil water content, BD (in g cm^{-3}) is the bulk density, WD (in g cm^{-3}) is the density of water, and PD is the particle density of the average soil material (quartz: 2.65 g cm^{-3}).

Bulk density, soil organic carbon (SOC, in kg C (kg soil)$^{-1}$) and soil pH were determined in the middle of the measurement period. To determine soil bulk density undisturbed soil samples were taken using stainless steel soil cores of known volume (100 cm^3). Then the samples were oven dried at 105°C for 24 h and determining the mass afterwards by weighing. SOC was determined by relating the mass difference of two soil sub-samples (5 g) – one air dried and the other dried at 430°C in a muffle furnace (until constant weight was achieved) – to the mass of the air dried sub-sample. For the determination of soil pH the soil was homogenized and afterwards measured in a soil-to-water suspension (1:2.5) using a glass electrode.

2.5 Calculation of the quality of simulation

We used the coefficient of determination (R^2), the model-efficiency (R^2eff), and the Root-Mean-Square Prediction Error (RMSE) to calculate the quality of the model

$$R^2 = \frac{(\Sigma(X_{mod}-\overline{X}_{mod})(X_{meas}-\overline{X}_{meas}))^2}{\Sigma(X_{mod}-\overline{X}_{mod})^2(X_{meas}-\overline{X}_{meas}))^2} \qquad (2)$$

$$R^2 eff = 1 - \left(\frac{\Sigma(X_{mod}-X_{meas})^2}{\Sigma(X_{meas}-\overline{X}_{meas})^2}\right) \quad (3)$$

$$RMSE = \sqrt{\frac{\Sigma(X_{mod}-X_{meas})^2}{n}} \quad (4)$$

where X_{mod} are the modelled N_2O fluxes and X_{meas} are the measured N_2O fluxes.

The coefficient of determination (R^2) is the most frequently used measure for the evaluation of the model quality. R^2 indicates the amount of the variance explained by the model. Thus the dispersion of the measuring data gives to be explained by the dispersion of the model data can at to which portion. The values can lie between 0 and 1, whereby values closed to 1 a good adjustment of the observation by the regression equation.

The R_2eff computes the measure for the relative deviation of the simulated values of the measured values under consideration of the dispersion of the measured values. The measure takes fast large negative values in the case of deviations of the model values to the measured values.

The RMSE (root mean square error) compares the model agreement with the average values of the measuring data.

2.6 Statistical analysis

A mean N_2O flux was calculated from the 15 plots for each day of N_2O flux measurement. These mean N_2O fluxes were tested by the t-test for significant departures from zero (confidence level 0.05).

Further, we tested whether correlations between N_2O fluxes at different plots depend on spatial distance between the plots. We first extracted the spatial coordinates of all 15 plots and computed their mutual distances dist (i,j) for all distinct pairs {i, j}, where i,j are chosen from {1, 2, ...15}, yielding 15 × 14/2 = 105 distances. For each of the 105 possible pairs of plots we also computed the cross-correlation coefficient C_{flux} (i,j) of their flux time series (34 time points each). The resulting correlation between the distances and flux correlations was tested

Appendix D

for significance using bootstrap permutation. We repeatedly associated the flux values to randomly chosen plots, again computed the cross-correlation coefficient C_{rand} (i,j), and generated a histogram of 20000 such randomly occurring correlations obtained from random association of actual fluxes to actual plot locations.

An analogous bootstrap analysis (based on 20000 random permutations) was performed for accessing the significance of correlations between soil parameters (soil pH, soil organic carbon content, clay content and bulk density) and N_2O fluxes.

3 Results

3.1 Field N_2O fluxes

Fig. 2 presents an overview of air temperature, precipitation and field N_2O fluxes from 15 plots distributed over the Hainich research site for the years 2005 and 2006. The grey dots display the uncertain N_2O fluxes (n=299) and the black dots the certain N_2O fluxes (n=211). However, the chance that the uncertain N_2O fluxes apply is higher than that the uncertain N_2O fluxes are zero. Also Fig. 3 shows that the mean N_2O fluxes calculated with the 211 certain N_2O fluxes (red dots) are in the range of the standard deviation of the N_2O fluxes calculated with all (n=510) N_2O fluxes (black dots). Except for the certain N_2O flux occurred at 6.September 2006. This N_2O flux exceeds clearly the N_2O flux calculated with all values. However, the certain N_2O flux consists only of one value because all other values measured at this date are uncertain. For this reason the N_2O flux occurred at 6.September 2006 is no mean N_2O flux and would not be considered in the further data evaluation. Therefore, we decided to calculate further results with all N_2O fluxes (n=510).

A seasonal pattern of field N_2O fluxes was lacking. The mean N_2O field fluxes are shown in Fig. 4. Average field N_2O fluxes (November 2005 to November 2006) exhibited small amplitudes between -5.03 and 38.76 µg m^{-2} h^{-1}, but most average field N_2O fluxes do not significantly differ from zero (t-test, p = 0.05) (see Fig. 3 and 4). 38% of all (n=510) observed N_2O fluxes show negative values. The highest field N_2O fluxes occurred between January and February 2006. During this time there was a frost period with soil temperature always below -0.5°C. This period contributes up to 40% to the observed field N_2O emission. A second period with mean field N_2O fluxes significantly different from zero started at the end of June 2006 (June 21, 2006 and June 28, 2006) and contributes up to 15% to the observed field N_2O emission. During this period the air temperature strongly increased (see Fig. 2). The annual field N_2O emission (derived from our field measurements) at the Hainich research site for the one year period (November 2005 to November 2006) was 0.46 ± 0.32 kg ha^{-1}.

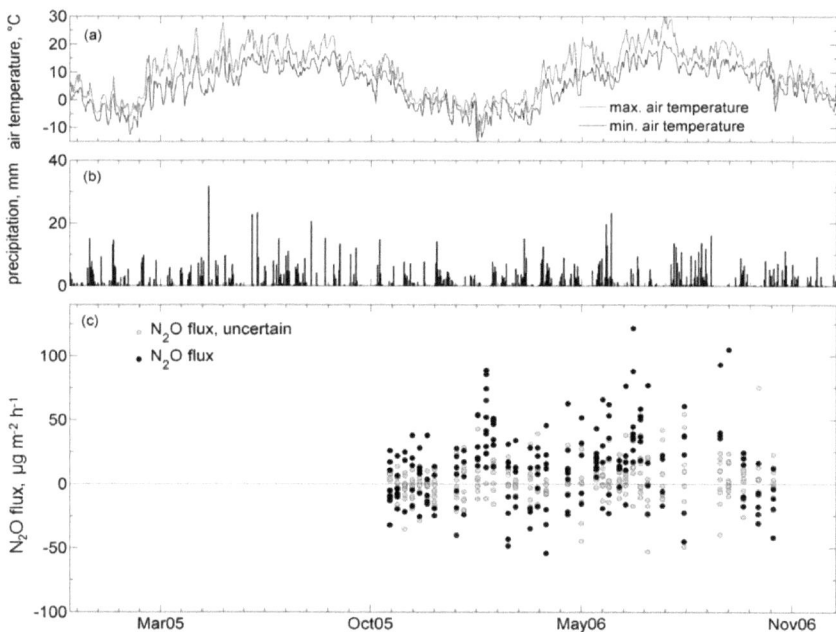

Figure 2: (a) Daily mean air temperature for 2005 and 2006 recorded 2 m above the ground and (b) daily precipitation for 2005 and 2006, (c) N_2O fluxes at the Hainich research site measured at 15 plots; black dots are the certain field N_2O fluxes and the grey dots display the uncertain N_2O fluxes. All N_2O fluxes are expressed in terms of mass of nitrogen.

Appendix D

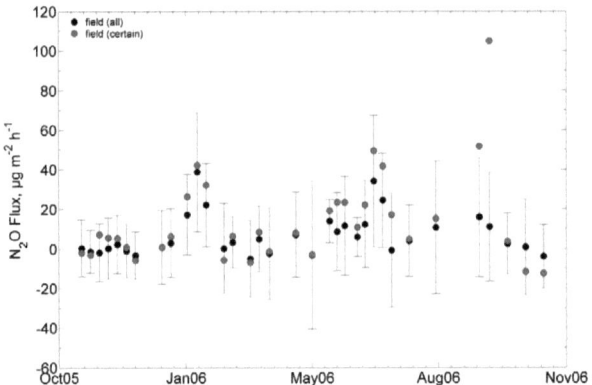

Figure 3: Mean field (all) and mean field (certain) N_2O fluxes for the Hainich research site form November 2005 to November 2006. The black dots are the mean field N_2O flux rates calculated with all values (n=15), the red dots are the mean field N_2O flux rates calculated with the certain values (see Fig.2). The error bars on each individual data point (only at the black dots) are the standard deviation. All N_2O fluxes are expressed in terms of mass of nitrogen.

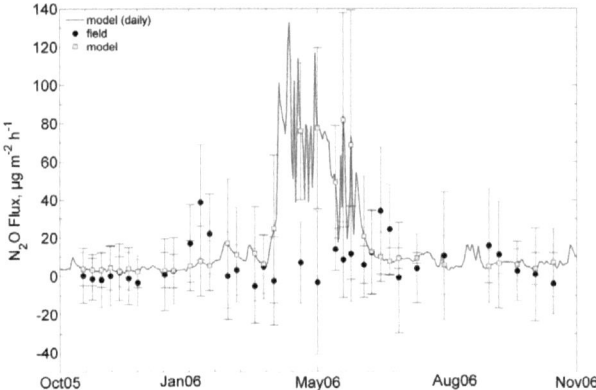

Figure 4: Field and modelled N_2O fluxes for the Hainich research site form November 2005 to November 2006. The black dots are the mean field N_2O flux rates (n=15), the grey squares are the mean modelled N_2O flux rates (n=15) and the grey line shows the mean daily modelled N_2O flux rates. The error bars on each individual data point are the standard deviation. All N_2O fluxes are expressed in terms of mass of nitrogen.

Appendix D

3.2 Soil climate

Fig. 5 presents the daily soil temperature and soil moisture simulated by Forest-DNDC and the mean field soil temperature and soil moisture measured at the time of gas sampling from plot 1. All other plots showed a similar pattern (not shown here). The modelled soil temperature matched well with the field soil temperature except for soil temperature near freezing (R^2=0.91, R^2eff=0.82, RMS=2.48). The modelled soil water content shows good agreement in the period from October 2005 to June 2006, but in the period from July to October 2006 the Forest-DNDC model overestimates the soil water content (Fig. 5). However, the model efficiency shows with 0.99 a very good agreement. The modelled soil water content showed no values below 0.57 WFPS. In contrast, the field soil water content ranged between 0.28 and 0.71 WFPS.

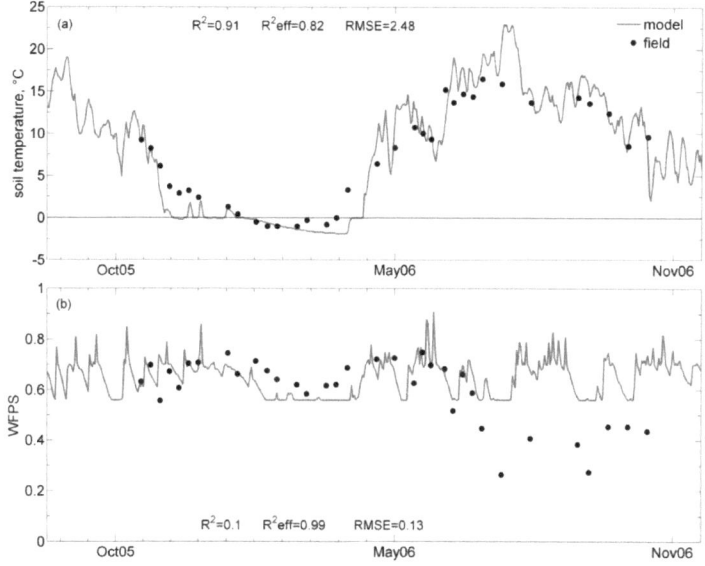

Figure 5: (a) Daily soil temperature (measured in 12 cm depth) from plot1 simulated by Forest-DNDC. Solid circles represent the soil temperature (integral between 0 and 12 cm) measured at the time of gas flux sampling at plot 1. (b) Daily soil WFPS (0-12 cm depth) from plot 1 simulated by Forest-DNDC. Solid circles represent the WFPS calculated with the gravimetric soil water content measured at the time of gas flux sampling at plot 1 (soil was taken from the first 10 cm).

Appendix D

3.3 Modelled N_2O fluxes

Mean modelled N_2O fluxes (derived from the mean of the modelled N_2O fluxes for each plot at the date of N_2O flux measurements) (2.54-81.72 µg m^{-2} h^{-1}, n=15) showed values typically larger than the plot averaged N_2O fluxes of the field measurements (-5.03-38.76 µg m^{-2} h^{-1}, n=15). The mean daily modelled N_2O fluxes (1.4-133.09 µg m^{-2} h^{-1}) were up to three times larger than the field N_2O fluxes. Mean modelled N_2O fluxes showed lowest values from November 2005 to March 2006 and from September to November 2006. Highest values occurred at the end of April 2006 (see Fig. 3).

Table 2: Overview of N_2O fluxes, range of mean N_2O fluxes and annual N_2O emission for field measurements, modelled N_2O fluxes of the 34 measuring days and modelled N_2O fluxes in daily resolution for the whole measuring period.

	range of mean N_2O fluxes, µg m^{-2} h^{-1}	annual N_2O emission, kg^{-1} ha^{-1}
field fluxes	-5.03-38.76	0.46±0.32
modelled fluxes (of the 34 measuring days)	2.54-81.72	1.77±1.82
modelled fluxes (daily resolution for the measuring period)	1.4-133.09	1.56±0.006

Furthermore, the daily modelled N_2O fluxes showed a weak seasonal pattern. The model does not account for N_2O uptake. The annual modelled N_2O emission at measurement intervals and the annual modelled N_2O emission in daily resolution for the Hainich research site for the one year measuring period (November 2005 to November 2006) were 1.77±1.82 and 1.56±0.006 kg ha^{-1}.

Appendix D

3.4 Quality of simulation

Table 3: Overview of the quality of simulation. Shown are the coefficient of determination (R^2), the root mean square (RMSE), the standard deviation (SD), the model efficiency (R^2eff) and the sample size (n).

Plot	R^2	RMSE ($\mu g\ m^{-2}\ h^{-1}$)	SD ($\mu g\ m^{-2}\ h^{-1}$)	R^2eff
1	0.001	40.5	22.2	-2.42
2	0.012	46.7	23.6	-3.04
3	0.074	33.8	30.8	-0.23
4	0.073	100.9	19.7	-25.97
5	0.016	32.5	20.6	-1.56
6	0.002	36.4	19.0	-2.78
7	0.001	39.7	24.1	-1.80
8	0.000	34.2	29.4	-0.40
9	0.021	29.9	24.5	-0.54
10	0.020	29.8	16.4	-2.41
11	0.040	22.06	17.8	-0.60
12	0.078	37.4	18.9	-3.05
13	0.000	23.2	20.8	-0.31
14	0.057	25.4	22.9	-0.35
15	0.122	37.0	27.1	-1.10

The results of the measures for the quality of simulation (see section 2.5) reflect the not good match between the measured and modelled N_2O fluxes. However, without the frost period and the period in summer modelled and field N_2O fluxes agree mostly within a factor of 2.

3.5 Spatial distribution of N_2O fluxes

The 510 individual measurements from the 15 plots resulted in field N_2O fluxes from -101.7 to 121.6 $\mu g\ m^{-2}\ h^{-1}$. During the frost period and the period end of June nearly all plots exhibited positive field N_2O fluxes (Fig. 2). The spatial variability showed high values in this period (range: -15.55 to 88.9 µg m^{-2} h^{-1}, mean: 26.04 µg m^{-2} h^{-1}). In the second period (end of June) with N_2O fluxes that were significantly different from zero, the spatial variability was also high (range: -17.21 to 121.6 µg m^{-2} h^{-1}, mean: 29.34 µg m^{-2} h^{-1}). During the periods with a background emission pattern (Brumme et al., 1999), both negative and positive N_2O fluxes occurred at similar ratio (range: -101.73 to 105.08 µg m^{-2} h^{-1}, mean: 3.36 µg m^{-2} h^{-1}). The modelled N_2O fluxes of the 15 individual plots showed no negative values. They ranged between 0.0 and 255.7 µg m^{-2} h^{-1} (see Fig. 6). The 15 individual plots exhibited the highest difference in modelled N_2O fluxes from each other (high standard deviation) on May, 23 2006 (Fig. 6). Plot 4, and, to a lesser degree, plot 9 and plot 10, with high N_2O fluxes, drive this high standard deviation.

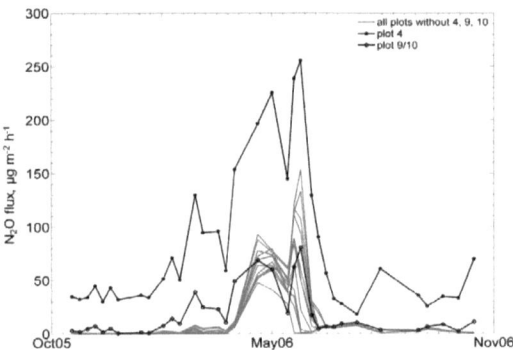

Figure 6: Individual modelled N_2O fluxes for all 15 plots of the Hainich research site. Plot 9 and 10 and plot 4 show different courses in comparison with the other plots.

3.6 Spatial correlation of N_2O fluxes

We find a weakly positive correlation between the correlation of N_2O fluxes at two plots and the distances between these plots (corr. Coefficient 0.085). This suggests that flux correlations between plots tend to be larger the more distant the plots (Fig.7). However, flux correlations are broadly distributed in the range between -0.3 and +0.77 and the best linear fit (least square regression) shows only a low average increase of flux correlations with distance at 0.17/km. Together with the small correlation coefficient, this suggests that the correlation between the

N₂O fluxes at any two plots may be almost independent of the distance between them. Indeed, a randomized (bootstrap) sampling of the given data (see Methods) yields some insight into whether this slightly positive correlation might be significant. We therefore created 20000 random permutation samples by randomly associating plots with flux time series, obtaining the distribution shown in Fig. 8b. We observe that 25.7% of the correlations between N_2O flux-correlations and randomized plot locations are larger than the observed value of 0.084 and 74.3% are lower. This strongly indicates that there is no relevant correlation between the actual inter-plot distances and the actual inter-plot N_2O flux correlations at spatial distances below 400 m.

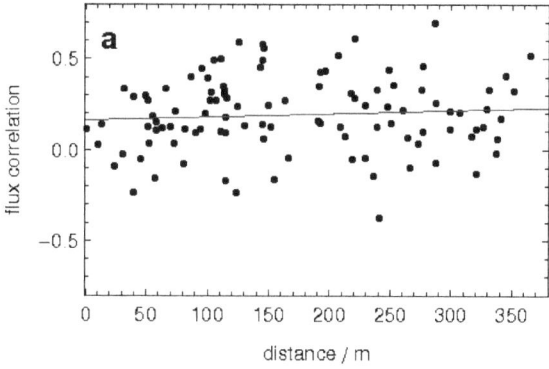

Figure 7: Inter-plot flux correlations do not significantly correlate with inter-plot distances. (a) Inter-plot correlations of nitrous oxide fluxes displayed versus inter-plot distances. The weakly positive trend is indicated by the least-squares linear fit (line, rate of change in correlation is 0.17/km).

3.7 Physical and chemical soil parameters

Physical and chemical soil parameters (soil pH, soil organic carbon (SOC), clay content, bulk density (BD)) at the 15 plots are summarized in Tab. 3. Although forest structure and land use history in the research area are homogeneous (Jordan et al., 2009), soil parameters are quite heterogeneous. Bulk density ranged from 0.79 g cm⁻³ (plot 2) to 0.92 g cm⁻³ (plot 4). Soil pH showed the lowest value at plot 2 (4.8) and the highest soil pH at plot 4 (6.7). The clay content and the soil organic carbon showed even larger variation. The clay content varied from 30.8% to 51% and the SOC ranged from 0.032 to 0.087 kg⁻¹ C kg soil. Apart from

Appendix D

significant correlations between bulk density and N_2O fluxes at six (out of 34) days of measuring (Fig. 7d), no significant correlations between N_2O fluxes and soil physical or chemical soil parameters were observed (Fig. 8).

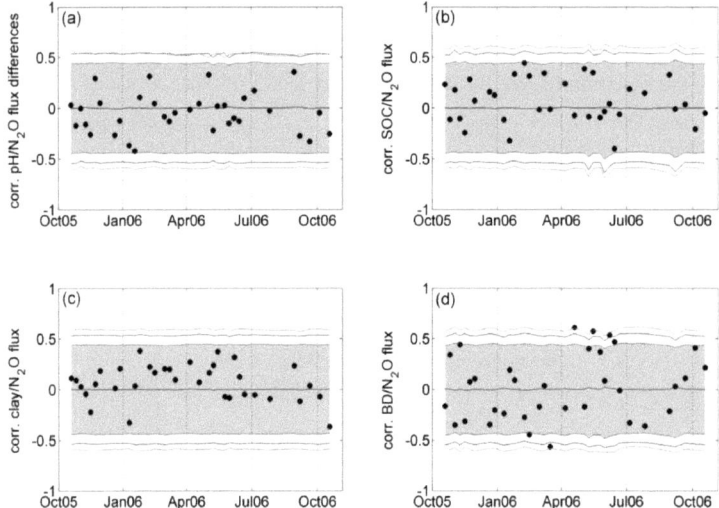

Figure 8: Correlations between fluxes and soil parameters. No significant correlations between soil parameters (a) pH value, (b) soil organic carbon content (SOC) and (c) clay content and fluxes. (d) Significant correlations between bulk densities and flux rates were observed only at six times of measurement. Significance tested by permutation test (20000 random site-permuted data samples, see methods). Red shaded area indicates mean plus and minus 45% whereas red, blue and green lines indicate 5%, 2%, and 1% chance levels for positive and negative correlations, respectively.

Table 3: Plot-specific soil parameters (soil pH, soil organic carbon (SOC), clay content and bulk density (BD)) of all 15 plots of the Hainich research site.

plot	pH [1]	SOC kg^{-1} C kg soil	clay %	BD g^{-1} cm^{3}
1	4.92	0.033	36.2	0.90
2	4.79	0.032	30.8	0.86
3	5.34	0.05	45.3	0.92
4	6.71	0.087	50.5	0.88
5	5.82	0.04	42.0	0.84
6	5.42	0.036	39.3	0.86
7	5.27	0.041	36.4	0.87
8	6.22	0.047	41.2	0.86
9	6.0	0.064	51.0	0.89
10	6.56	0.062	48.6	0.79
11	5.74	0.052	44.2	0.79
12	5.28	0.054	35.0	0.81
13	6.37	0.05	44.5	0.86
14	6.14	0.041	38.7	0.87
15	6.24	0.046	40.2	0.91

4 Discussion

4.1 Comparison with other studies

All previous studies focusing on N_2O fluxes from soils in beech forest ecosystems were performed in managed forest ecosystems (Brumme et al., 1999; Papen and Butterbach-Bahl, 1999; Butterbach-Bahl et al., 2002; Gasche and Papen, 2002; Brumme and Borken, 2009; Borken and Beese, 2006; Zechmeister-Boltenstern et al., 2002), whereas we measured and studied N_2O fluxes in a beech forest that is long-term unmanaged. Still our results are consistent with the results of Brumme et al. (1999) and Brumme and Borken (2009) for forests dominated by the background emission type. They found that soils in beech forests with a mull organic horizon usually show background emissions sometimes interrupted by event emissions like frost and thaw. Our annual N_2O emission (0.46 ± 0.32 kg ha^{-1}) was within the lower range of values reported for temperate beech forest soils (see Tab. 4).

Contrasting to the other studies, the core zone of the Hainich National Park is a close to natural deciduous forest which is unique in Central Europe. This could be a reason why most N_2O fluxes did not significantly differ from zero and the overall mean flux is particularly low. It may also well be that N is immobilized as a part of rising soil organic matter stocks (Gleixner et al., 2009). In any case, our results clearly support the hypotheses that most natural ecosystems do not emit significant amounts of N_2O and natural background emissions are rather low. Guckland (2009) found even lower emissions of N_2O ranging from -31.4 to 167.8 µg m^{-2} h^{-1} at a site covered to 59% with beech also located in the Hainich NP for 2005 to 2007. These measurements were performed on loess soils (Luvisol) with less clay content than in soils studied here resulting in an annual N_2O emission of 0.19 ± 0.16 kg ha^{-1}. The N_2O fluxes measured by Guckland (2009) during freezing and thawing amounted to 94% of the emissions of the first year. Our values of 38% are closer to the values observed by Papen and Butterbach-Bahl (1999) who estimated a contribution up to 39% to the total annual N_2O emission caused by freezing and thawing at a 96 year-old beech plantation in the Höglwald (Germany) (see also Tab. 4). Therefore, the magnitude of N_2O emissions at the Hainich site took place during frost. Similar results were also observed by Butterbach-Bahl et al. (2002), Papen and Butterbach-Bahl (1999), Brumme et al. (1999) and Teepe et al. (2004) which also found that the magnitude of N_2O emission take place during frost. A peak of N_2O emission during thawing could not be observed. However, this might be due to the weekly to biweekly resolution of N_2O flux measurements. Generally, fluxes may be underestimated by low temporal resolution of measurements during short periods of extremely high N_2O fluxes like frost-thaw events (Flessa et al., 2002a). However, we observed no seasonal pattern of N_2O fluxes which would be detectable with our time measurement resolution and is in accordance with the results from Brumme et al. (1999) and Brumme and Borken (2009) who also reported no occurrence of seasonal pattern by forests with N_2O emissions <1 kg ha^{-1} yr^{-1}.

Borken and Beese (2006) reported annual N_2O emission of 0.54 ± 0.14 kg ha^{-1} and 1.55 ± 1.8 kg ha^{-1} for the years 1999-2000 and 2000-2001 for a pure beech site at Solling (Germany).

Table 4: Compilation of published annual N_2O fluxes from soils of temperate beech forest ecosystems (N_2O fluxes are in terms of mass of nitrogen)

Site	N_2O emission, kg ha^{-1} yr^{-1}	Observation period	Reference
Hainich[a], Germany	0.46 ± 0.32	2005-2006	this study
Hainich[b], Germany	0.19 ± 0.16	2005-2007	Guckland (2009)
Göttinger Wald, Germany	0.17 ± 0.03	1993-1995	Brumme and Borken (2009)
Zierenberg, Germany	0.41 ± 0.12	1991-1992	Brumme and Borken (2009)
Solling, Germany	1.93 ± 0.63	1990-2000	Brumme and Borken (2009)
Solling, Germany	0.54 ± 0.14	2000-2001	Borken and Beese (2006)
Schottenwald, Austria	4.03 ± 1.37	1996-1998	Zechmeister-Boltenstern et al. (2002)
Höglwald, Germany	5.1	1995-1996	Papen and Butterbach-Bahl (1999)
Höglwald, Germany	3.8 ± 1.3	1995-1997	Butterbach-Bahl et al. (2002)

[a] main soil substrate: clay [b] main soil substrate: loss

Significantly higher annual N_2O emissions usually observed for sites showing a clear seasonal pattern with low N_2O fluxes during the winter months and high N_2O fluxes during the summer. Seasonal N_2O flux pattern were observed by Zechmeister-Boltenstern et al. (2002) for a 140 year-old beech site located at Schottenwald (Austria); they reported annual emissions of 3.6 ± 1.0 kg ha^{-1} (1996), 4.2 ± 1.3 kg ha^{-1} (1997) and 4.3 ± 1.8 kg ha^{-1} (1998). Kitzler et al. (2006) found N_2O fluxes ranging between -6.3 and 75.4 µg m^{-2} h^{-1} for a 142 year-old beech site at Schottenwald (Austria) (observed year: 2002). Seasonal variations of

Appendix D

N$_2$O fluxes at this site followed mainly the annual changes in soil temperature, soil moisture and the availability of nitrogen in the soil.

4.2 Measured vs. modelled N$_2$O fluxes

Our results show that field and modelled soil temperature fit well except for the period with soil temperature near the freezing point. The soil cooled down slower and warmed up faster than the model simulation. Similar results were found by Szyska et al. (2008). In contrast, field soil moisture differed significantly from modelled soil moisture, which does not fall below 0.57 WFPS. The reason for this is the model settings, particularly the wilting point, for the default soil type "clay loam". However, both parameters significantly affect the N$_2$O fluxes (Pathak, 1999; Saggar et al., 2004; Smith et al., 2003) and Forest-DNDC was designed with a strong soil moisture control on N$_2$O fluxes (Frolking et al., 1998; Saggar et al., 2004; Giltrap et al., 2009; Stange et al., 2000). Only a few studies report about the match between measured and simulated (by Forest-DNDC) soil water contents. Kröbel et al. (2010) found an overestimation of the soil water content by DNDC (version DNDC89). At a Scottish site on a glacial till with very low hydraulic conductivity in the subsoil which makes drainage very slow, Frolking et al. (1998) reported DNDC to underestimate soil moisture. Beheydt et al. (2007) also mentioned an underestimation of WFPS for different investigated sites in their study using the DNDC version 8.3P. Saggar et al. (2004) reported a poor match between measured and simulated WFPS for a silt loam soil in New Zealand using NZ-DNDC. The simulation had low WFPS when the topsoil was almost saturated with water and higher WFPS when the topsoil was dry. Stange et al. (2000) observed a underestimation of WFPS during winter and spring and an overestimation of WFPS by PnET-N-DNDC in summer and, especially, in early autumn for a beech site in the Höglwald, Germany. This discrepancy between measured and simulated values of WFPS at the beech site is mainly due to an underestimation of evapotranspiration by the beech stand. However, all processes related to C- and N-cycling in forest ecosystems are strongly dependent on water availability of soil. Therefore, a successful simulation of soil hydrology in forest ecosystems is necessary in order to exactly model C- and N-cycling in forest soil (Saggar et al., 2004; Stange et al., 2000).

The annual N$_2$O emission of the Forest-DNDC simulation overestimated the annual emission derived from our field measurements. Also Abdalla et al. (2009) found that DNDC poorly

Appendix D

described those fluxes from zero fertilizer treatments. Forest-DNDC also fails to correctly simulate the actual fluctuations of measured N_2O fluxes except for the plot with the highest SOC content. Plot 4 reflected a part of the fluctuations. Forest-DNDC failed to show the N_2O fluxes during frost except for plot4, plot9, and plot10. These plots showed higher N_2O fluxes during frost. The model seems to be very sensitive to the input variable, soil organic carbon (SOC), because all three plots exhibit SOC contents >0.06 kg^{-1} C kg soil. At the plots with less than 0.06 kg^{-1} C kg soil SOC, no increasing N_2O fluxes during frost took place. The model shows thaw induced elevated N_2O fluxes, at the end of April. This event contributes to a great part to the annual N_2O emission. However, we did not observe a thawing peak in the field. Frolking et al. (1998) also reported disagreements between modelled N_2O fluxes and field N_2O fluxes especially during freezing and thawing. However, the second N_2O peak in summer 2006 did not appear in the simulated data. The simulated N_2O fluxes decreased in this period. Saggar et al. (2004) remarked that the DNDC model had limited success in predicting the size and timing of very high fluxes. It seems to be impossible to simulate the correct temporal variability of N_2O fluxes. However, these results show that this model is not yet able to display the lack of knowledge in this area and not all controls of N_2O fluxes seems to be well understood.

A further point is that modelled N_2O fluxes exhibited no negative fluxes. 38% of the field N_2O fluxes were negative and also other studies observed negative N_2O fluxes (Kitzler et al., 2006; Chapuis-Lardy et al., 2007; Guckland, 2009). However, the field N_2O fluxes are usually small and the standard errors of these fluxes are high. Also a simulation of N_2O uptake by forest soils with the DNDC version presented by Li et al. (2000) was not possible (Stange, 2001). However, the PnET-N-DNDC version presented by Stange (2001) simulated N_2O uptake. Theoretically Forest-DNDC can simulate N_2O uptake, but the parameter settings for denitrification in uncalibrated versions of Forest-DNDC do not permit N_2O uptake. To allow N_2O uptake a new parameterization must be included (personal communication, Butterbach-Bahl, 2010).

4.3 Spatial variability of N_2O fluxes

Folorunso and Rolston (1984) reported that spatial variability in N_2O fluxes is naturally large in most soils. That was confirmed by our data. Our field N_2O fluxes observed at the 15 individual plots show their highest spatial variability when high N_2O fluxes occur. However,

Appendix D

also in periods with low mean N_2O fluxes the variability was pronounced. The modelled N_2O fluxes simulated with individual soil parameter from the 15 plots also exhibited a high spatial variability. This is due to the differences in the parameter settings for SOC. N_2O fluxes at plots with low (0.032-0.054 kg^{-1} C kg soil) SOC content fluctuated less than plots with high (0.062-0.087 kg^{-1} C kg soil) SOC concentrations. The field N_2O fluxes did not display this effect. Nevertheless, also the modelled N_2O fluxes showed their highest variability when high N_2O fluxes occurs.

4.4 Spatial correlation of N_2O fluxes

One may assume that N_2O fluxes from nearby plots are more strongly correlated than N_2O fluxes from more distant plots, but we are not aware of any studies on this issue. Studies about spatial variability usually focus on differences of N_2O fluxes caused by differences in soil properties (Ambus and Christensen, 1995; Röver et al., 1999). For our 15 different plots at the Hainich research site we found no consistent significant dependence of the N_2O fluxes on soil parameters.

We observed no significant spatial correlation of N_2O fluxes at the Hainich research site. Although we found a weak tendency towards a positive correlation, this does not explain the spatial variability in N_2O fluxes by changes in soil properties. However, Mummey et al. (1997) explained the variation of N_2O fluxes by availability of soil nutrients and Ambus and Christensen (1994) reported a strong correlation by plots less than 1 m apart, supposedly due to soil aeration patterns influenced by the dispersion of anaerobic microsites.

5 Conclusions

In this study we investigated field and modelled N_2O fluxes for the first time from an unmanaged old growth beech forest in Central Europe (Hainich National Park, Germany).

Our results reveal particularly low N_2O emissions comparable to the lowest observed in managed beech forests. They clearly underline that natural background emissions from this ecozone are lower than 1 kg N ha^{-1} yr^{-1}. Furthermore, the results indicate that site properties other than forest management control substantially affect the magnitude of N_2O emissions.

The absence of spatial correlations indicates that within one site the distance between each chamber is secondary. This is valuable for designing measurement plots because larger distances between individual chambers are not required. The latter of course, needs to be further verified by additional studies. Our study also indicates that an uncalibrated Forest-DNDC model is not fully appropriate for simulating annual fluxes of N_2O for zero fertilizer treatments, if it is the aim to obtain results for more than one particular site. Therefore, regionalization with such a biogeochemical mechanistic model that has inevitably to be "uncalibrated" appears inappropriate. To simulate the temporal variability in N_2O fluxes, a validation at other research sites seems to be necessary especially for sites with low soil organic carbon values. Therefore, it remains a challenge for future research to satisfactorily reproduce the spatial variability of natural N_2O fluxes.

Some other, managed forest ecosystems with absent seasonally or event-induced N_2O emissoins also display annual N_2O emission at or below 0.5 kg ha^{-1} yr^{-1}. Hence a background flux around 0.5 kg ha^{-1} yr^{-1} for the given background emission type and the given background emission factor seems to be more adequate.

Acknowledgements

Our work was funded by the Deutsche Forschungsgemeinschaft (Gl 308/5-1 and 5-2). We acknowledge the help of Petr Holy in collecting gas samples and the Max Planck Institute of Biogeochemistry for providing us the climate data.

Appendix D

References

Ambus, P., and Christensen, S.: Measurement of N_2O emission from a fertilized grassland: An analysis of spatial variability, J Geophys Res, 99, 1994.

Ambus, P., and Christensen, S.: Spatial and Seasonal Nitrous Oxide and Methane Fluxes in Danish Forest-, Grassland-, and Agroecosystems, J Environ Qual, 24, 993-1001, 1995.

Beheydt, D., Boeckx, P., Sleutel, S., Li, C., and Van Cleemput, O.: Validation of DNDC for 22 long-term N_2O field emission measurements, Atmos Environ, 41, 6196-6211, 2007.

Borken, W., and Beese, F.: Methane and nitrous oxide fluxes of soils in pire and mixed stands of European beech and Norway spruce Eur J Soil Sci, 57, 617-625, 10.1111/j.1365-2389.2005.00752.x, 2006.

Bouwman, A. F.: Direct emission of nitrous oxide from agricultural soils, Nutr Cycl Agroecosys, 46, 53-70, 1996.

Brumme, R., Borken, W., and Finke, S.: Hierarchical Control on Nitrous Oxide Emission in Forest Ecosystems, Global Biogeochem. Cycles, 13, 1137-1148, 10.1029/1999gb900017, 1999.

Brumme, R., and Borken, W.: N_2O Emission from Temperate Beech Forest Soils, in: Functioning and Management of European Beech Ecosystems, edited by: Brumme, R., and Khanna, P. K., Springer, Berlin, 353-367, 2009.

Butterbach-Bahl, K., Stange, F., and Papen, H.: Regional inventory of nitric oxide and nitrous oxide emissions for forest soils of southeast Germany using the biogeochmical model PnET-N-DNDC, J Geophys Res, 106, 34,155-134,166, 2001.

Butterbach-Bahl, K., Gasche, R., Willibald, G., and Papen, H.: Exchange of N-gases at the Höglwald Forest – A summary, Plant Soil, 240, 117-123, 2002.

Chapuis-Lardy, L., Wrage, N., Metay, A., Chotte, J.-L., and Bernoux, M.: Soils, a sink for N_2O? A review, Glob Change Biol, 13, 1-17, 2007.

Ellenberg, H.: Vegetation Mitteleuropas mit den Alpen aus ökologischer, dynamischer und historischer Sicht, in, Ulmer, Stuttgart, 1996.

Flessa, H., Ruser, R., Dörsch, P., Kamp, T., Jimenez, M. A., Munch, J. C., and Beese, F.: Integrated evaluation of greenhouse gas emissions (CO_2, CH_4, N_2O) from two farming systems in southern Germany, Agriculture, Ecosystems & Environment, 91, 175-189, 2002a.

Flessa, H., Ruser, R., Schilling, R., Loftfield, N., Munch, J. C., Kaiser, E. A., and Beese, F.: N_2O and CH_4 fluxes in potato fields: automated measurement, management effects and temporal variation, Geoderma, 105, 307-325, 2002b.

Folorunso, O. A., and Rolston, D. E.: Spatial Variability of Field-Measured Denitrification Gas Fluxes, Soil Sci Soc Am J, 48, 1214-1219, 1984.

Frolking, S. E., Mosier, A. R., Ojima, D. S., Li, C., Parton, W. J., Potter, C. S., Priesack, E., Stenger, R., Haberbosch, C., Dörsch, P., Flessa, H., and Smith, K. A.: Comparison of N_2O emissions from soils at three temperate agricultural sites: simulations of year-round measurements by four models, Nutr Cycl Agroecosys, 52, 77-105, 1998.

Gasche, R., and Papen, H.: Spatial variability of NO and NO_2 flux rates from soil of spruce and beech forest ecosystems, Plant Soil, 240, 67-76, 2002.

Giltrap, D. L., Li, C., and Saggar, S.: DNDC: A process-based model of greenhouse gas fluxes from agricultural soils, Agriculture, Ecosystems & Environment, In Press, Corrected Proof, 2009.

Gleixner, G., Tefs, C., Jordan, A., Hammer, M., Wirth, C., Nueske, A., Telz, A., Schmidt, U. E., and Glatzel, S.: Soil Carbon Accumulation in Old-Growth Forests, in, 231-266, 2009.

Groombridge, B., and Jenkins, M. D.: Global biodiversity. Earth's living resources in the 21st century., World Conservation Press, Cambridge, 2000.

Guckland, A.: Nutrient stocks, acidity, processes of N transformation and net uptake of methane in soils of a temperate deciduous forest with diefferent abundance of beech, PhD, Fakultät für Forstwissenschaften und Waldökologie, Georg-August-Universität Göttingen, Göttingen, 121 pp., 2009.

Jordan, A., Jurasinski, G., and Glatzel, S.: Small scale spatial heterogeneity of soil respiration in an old growth temperate deciduous forest, Biogeosciences Discussions, 6, 9977-100005, 2009.

Kitzler, B., Zechmeister-Boltenstern, S., Holtermann, C., Skiba, U., and Butterbach-Bahl, K.: Nitrogen oxides emission from two beech forests subjected to different nitrogen loads, Biogeosciences, 3, 293-310, 2006.

Kröbel, R., Sun, Q., Ingwersen, J., Chen, X., Zhang, F., Müller, T., and Römheld, V.: Modelling water dynamics with DNDC and DAISY in a soil of the North China Plain: A comparative study, Environmental Modelling & Software, 25, 583-601, 2010.

Li, C., Aber, J., Stange, F., Butterbach-Bahl, K., and Papen, H.: A process-oriented model of N_2O and NO emissions from forest soils: 1. Model development, J Geophys Res, 105, 2000.

Loftfield, N., Flessa, H., Augustin, J., and Beese, F.: Automated Gas Chromatographic System for Rapid Analysis of the Atmospheric Trace Gases Methane, Carbon Dioxide, and Nitrous Oxide, J Environ Qual, 26, 560-564, 1997.

Mosier, A. R., Duxbury, J. M., Freney, J. R., Heinemeyer, O., and Minami, K.: Nitrous oxide emissions from agricultural fields: Assessment, measurement and mitigation, Plant Soil, 181, 95-108, 1996.

Mummey, D. L., Smith, J. L., and Bolton, H.: Small-scale spatial and temporal variability of N_2O flux from a shrub-steppe ecosystem, Soil Biol Biochem, 29, 1699-1706, 1997.

Papen, H., and Butterbach-Bahl, K.: A 3-year continuous record of nitrogen trace gas fluxes from untreated and limed soil of a N-saturated spruce and beech forest ecosystem in Germany 1. N_2O emissions, J. Geophys. Res., 104, 10.1029/1999jd900293, 1999.

Parton, W. J., Holland, E. A., Del Grosso, S. J., Hartman, M. D., Martin, R. E., Mosier, A. R., Ojima, D. S., and Schimel, D. S.: Generalized model for NO_x and N_2O emissions from soils, J Geophys Res-Atmos, 106, 17403-17419, 2001.

Pathak, H.: Emissions of nitrous oxide from soil, Current Science, 77, 359-369, 1999.

Röver, M., Heinemeyer, O., Munch, J. C., and Kaiser, E.-A.: Spatial heterogeneity within the plough layer: high variability of N_2O emission rates, Soil Biol Biochem, 31, 167-173, 1999.

Ruser, R., Schilling, R., Steindl, H., Flessa, H., and Beese, F.: Soil Compaction and Fertilization Effects on Nitrous Oxide and Methane Fluxes in Potato Fields, Soil Sci Soc Am J, 62, 1587-1595, 1998.

Saggar, S., Andrew, R. M., Tate, K. R., Hedley, C. B., Rodda, N. J., and Townsend, J. A.: Modelling nitrous oxide emissions from dairy-grazed pastures, Nutr Cycl Agroecosys, 68, 243-255, 2004.

Skiba, U., and Smith, K. A.: The control of nitrous oxide emissions from agricultural and natural soils, Chemosphere - Global Change Science, 2, 379-386, 2000.

Smith, K. A., Ball, T., Conen, F., Dobbie, K. E., Massheder, J., and Rey, A.: Exchange of greenhouse gases between soil and atmosphere: interactions of soil physical factors and biological processes, Eur J Soil Sci, 54, 779-791, 2003.

Stange, F., Butterbach-Bahl, K., Papen, H., Zechmeister-Boltenstern, S., Li, C., and Aber, J.: A process-oriented model of N_2O and NO emissions from forest soils 2. Sensitivity analysis and validation, J Geophys Res, 105, 2000.

Stange, F. C.: Entwicklung und Anwendung eines prozeßorientierten Modells zur Beschreibung der N_2O- und NO-Emissionen aus Böden temperater Wälder, Ph.D thesis, Frauenhofer Institute for Atmosperic Environmental, Garmisch-Partenkirchen, Germany, 143 pp., 2001.

Szyska, B., Bach, M., Breuer, L., Frede, H.-G., and Vache´, K. B.: Indikatoren für eine nachhaltige Landnutzung - Modellkopplung zur Abschätzung von N-Emissionen aus der Pflanzenproduktion, Beiträge zur Hydrogeologie, 56, 150-158, 2008.

Teepe, R., Brumme, R., Beese, F., and Ludwig, B.: Nitrous Oxide Emission and Methane Consumption Following Compaction of Forest Soils, Soil Sci Soc Am J, 68, 605-611, 2004.

Zechmeister-Boltenstern, S., Hahn, M., Meger, S., and Jandl, R.: Nitrous oxide emissions and nitrate leaching in relation to microbial biomass dynamics in a beech forest soil, Soil Biol Biochem, 34, 823-832, 2002.

Die VDM Verlagsservicegesellschaft sucht für wissenschaftliche Verlage abgeschlossene und herausragende

Dissertationen, Habilitationen, Diplomarbeiten, Master Theses, Magisterarbeiten usw.

für die kostenlose Publikation als Fachbuch.

Sie verfügen über eine Arbeit, die hohen inhaltlichen und formalen Ansprüchen genügt, und haben Interesse an einer honorarvergüteten Publikation?

Dann senden Sie bitte erste Informationen über sich und Ihre Arbeit per Email an *info@vdm-vsg.de*.

Sie erhalten kurzfristig unser Feedback!

VDM Verlagsservicegesellschaft mbH
Dudweiler Landstr. 99   Telefon +49 681 3720 174
D - 66123 Saarbrücken   Fax   +49 681 3720 1749
www.vdm-vsg.de

Die VDM Verlagsservicegesellschaft mbH vertritt

Printed by Books on Demand GmbH, Norderstedt / Germany